Aristotle's *Μετεωρολογικά*

Aristotle's *Μετεωρολογικά*

Meteorology then and now

Anastasios A. Tsonis and Christos Zerefos

Archaeopress Publishing Ltd
Summertown Pavilion
18-24 Middle Way
Summertown
Oxford OX2 7LG

www.archaeopress.com

ISBN 978-1-78969-637-0
ISBN 978-1-78969-638-7 (e-Pdf)

© Anastasios A. Tsonis, Christos Zerefos, Mariolopoulos-Kanaginis Foundation for the Environmental Sciences and Archaeopress 2020

Cover image: Mt. Olympus summit at 3.3 km height, a few hundred kilometers from Aristotle's birthplace at Stagira (courtesy of Stelios Zerefos)

This book is available direct from Archaeopress or from our website www.archaeopress.com

 This work is licensed under a Creative Commons Attribution-NonCommercial-NoDerivatives 4.0 International License

Contents

List of figures iii
Prolegomena 1
Introduction: about Aristotle 5
    His life 5
    His works 6

BOOK A FROM ΜΕΤΕΩΡΟΛΟΓΙΚΑ

Aristotle's universe with a glimpse on climate change 9
    Meteorology now, part 1 19
Analogies and contrasts 32
    Key points of meteorology now, part 1 34
Back to Aristotle's *Meteorologica* 36
    Meteorology now, part 2 41
    Key points of meteorology now, part 2 52
Analogies and contrasts 53
Back to Aristotle's *Meteorologica* 56

BOOK B FROM ΜΕΤΕΩΡΟΛΟΓΙΚΑ

On winds 63
Stormy weather 74
    Meteorology now, part 3 76
    Key points of meteorology now, part 3 81
Back to Aristotle's *Meteorologica* 81

BOOK C FROM ΜΕΤΕΩΡΟΛΟΓΙΚΑ

Aristotle's optics 85
    Preparatory introduction 85
    Aristotle's general theory of colour 86
    The halo 89
    Rainbow 92
    Sun dogs and light pillars 98
    Aurora Borealis 99

BOOK D FROM ΜΕΤΕΩΡΟΛΟΓΙΚΑ

Aristotle's notion on thermodynamic equilibrium 105
Concluding remarks 107
Appendix I Aristotle on climate change 109
Appendix II On the meaning of the word virtue and Aristotle's poem 'Ode to Virtue' 117
Index 121

List of figures

Figure 1: Aristotle's view of the universe lasted for 2,000 years and had far reaching consequences. From the edition by Thomas Digges of his father's A Prognostication everlasting..., published in 1576 in London (by permission of the Royal Society). The solid lines define the spheres and the broken lines their motion (orbit)... 10

Figure 2: The relationship between the principles and the four elements. The corners of the inner square correspond to the four principles and the corners of the surrounding square to the four elements. 12

Figure 3: A simplified version of Aristotle's universe and the distribution of the five elements (aether, fire, air, water, and earth). The red arrows indicate the rotation of the upper world (heavenly bodies). 15

Figure 4: Ptolemy's model to explain the retrograde motion of planets.. 18

Figure 5: Distribution of particles with height. Because of gravity the atmosphere is denser close to the surface. As we go higher density decreases. 21

Figure 6: A simple demonstration on the relation of pressure and density. 23

Figure 7: A simple example demonstrating how rising air will create convection.................................. 27

Figure 8: The Hadley circulation cells 29

Figure 9: As the parcel of air rises it expands and eventually cools (see text for details). 31

Figure 10: The vertical structure of the atmosphere. The yellow line shows how temperature changes with height (courtesy of NOAA)... 33

Figure 11: As the parcel rises it cools and its relative humidity increases. 44

Figure 12: An illustration of how a cloud grows when the air that has reached saturation rises. 46

Figure 13: Illustration of the ice-crystal process. 48

Figure 14: The basic geometry of an ice crystal.. 49

Figure 15: A snowflake. 50

Figure 16: All or some of the snow formed in the upper cold layer may melt if it travels through a middle, melting layer. Then some raindrops may freeze going through the lower freezing layer thereby bringing a mixture of types of precipitation on the ground. 51

Figure 17: The three general wind belts at the surface of the planet. 68

Figure 18: The setup behind the etesians. The red colour shows elevation above 700 meters, which clearly forms a 'channel' by the mountains between Greece and Anatolia. The black lines show the mean surface pressure field for July-August, and the black arrow resulted direction of the wind (see text for details). 71

Figure 19: A simple diagram depicting the habitable zones (ABCD and EFGH) according to Aristotle. ..72

Figure 20: Population density on Earth. The grey horizontal lines are the Tropic of Cancer and the arctic circle, respectively. The darker the red colour the higher the density. Source: coolgeography.co.uk 73

Figure 21: Charge distribution with a thunderstorm... 78

Figure 22: Stages in the formation of a tornado. 80

Figure 23: Aristotle's explanation of why halos are circular. O stands for observer and S for Sun (or moon). 90

Figure 25: Illustration of halo formation. See text for details. 91

Figure 24: Refraction of light by an ice crystal. 91

Figure 26: Aristotle's set up and geometric illustration of rainbow production. S is the Sun, O is the observer at the centre of a sphere. The cloud is in front of the observer. 93

Figure 27: The spectrum of the visible light. 94

Figure 28: As light from the Sun enters a raindrop it undergoes refraction, reflection and refraction again, a process that splits it into its colour components. 94

Figure 29: Formation of rainbow. 95

Figure 30: Formation of the secondary rainbow. 96

Figure 31: Aristotle's law of reflection. Reflection to H from K can only occur from point M.................. 97

Figure 33: Halo around the Sun, sun dogs on the right and left of the Sun, and a light pillar above the Sun. Courtesy of Wikimedia Commons. Attribution Gopherboy6956..................................... 100

Figure 34: Aurora Borealis as seen from space (top) and from surface (bottom). Courtesy of NASA. ... 102

Prolegomena

Meteorology derives from the Ancient Greek word μετέωρος *metéōros* (*meteor*) which means 'things suspended high in the air'. Μετεωρολογικά (*Meteorologica*; on Meteorology) occupy a special place within the body of work by Aristotle. It is the only book dealing with many diverse areas such as astronomy, geometry, optics, geography, seismology, volcanology, chemistry, and today's aim of meteorology, weather forecasting and is divided into four books. Some of the features that Aristotle discusses in *Meteorologica* (for example, rivers and the seas), cannot be considered as suspended in the air. They are included in this book by virtue of their relation to moist and cold exhalation defined in Chapter 4 of Book A. Similarly, discussion on shooting stars, comets, and Milky Way has nothing to do with weather and meteorology. They are part of his discussion because they were thought as 'suspended' and products of the exhalation from the Earth. The four books are divided as follows:

Book A starts by considering the five fundamental elements: aether, fire, air, water, and earth, and their distribution in Aristotle's model of the universe. A supreme notion is then introduced, which he named αναθυμιαση or exhalation, which causes water and land to exhale vapour and dry air, respectively, after they are warmed by the Sun. The planet is warmed by solar radiation and then the air in contact with it warms up and rises. Aristotle then continues with a discussion about this rising motion and how it produces dew, frost, rain, snow, and hail. After that, the discussion moves to issues not directly related to weather such as stars, comets, the galaxy, etc. Aristotle discusses the phenomena of Aurora Borealis, which will be discussed in detail later in Book C, together with other optical phenomena such as rainbows. At the end of Book A, Aristotle touches on an important issue related to what today we would call global climate change, a topic of great interest and debate now-days. In this book, Aristotle also briefly talks about winds, a topic that forms a major part of Book B with the detailed elaboration on wind formation and wind distribution on Earth. In Book B, Aristotle continues with the formation of lightning and thunder and subsequently touches on earthquakes, a topic again not much related to meteorology. In Book C,

Aristotle discusses in detail optical phenomena such as rainbow and halo where with the help of geometry he explains their formation. In this book, Aristotle also ventures into the creation of metals, which again is not much related to meteorology. In Book D, Aristotle classifies 'warm' and 'cold' as the two 'active' causes and 'dry' and 'moist' as the two 'passive' causes, and goes on to discuss how the active affect the passive and how their interaction affects our senses. This book is hardly related to meteorology; however, it appears that hidden in this discussion is the notion of (thermodynamic) equilibrium that is important in weather processes.

It is interesting to note that Aristotle did not provided titles for his works. The titles, including Μετεωρολογικά, were invented later for convenience by his publishers. The chronology of Μετεωρολογικά is not very clear, as it is one of his works that refers to very few historical events at the time of the writing. One such event is the fire of the temple of Ephesus in 356 BC. Another is the appearance of a comet in Athens in the time of ruler Nikomachus (around 341-340 BC). These two events suggest an interval of 16 years (356-340 BC) in which Μετεωρολογικά were written. In addition, it is worth mentioning that while the authenticity of the first three books is undisputed, some scholars have raised questions about the authenticity of the fourth book; however, most of the scholar agree that it is definitely a work of Aristotle, and that differences in style probably reflect the fact that in writing Book D, Aristotle used notes from his teaching students at the Academy.

While there are scientific issues behind Aristotle's writings, this book is written for non-specialist. We use simple examples which will be easily followed by general readers.

In this study we will consider only parts relevant to meteorological phenomena (for example, we will not deal with Aristotle's views on the stars or earthquakes) and we will compare Aristotle's arguments with the current knowledge of meteorology. The purpose of this book, however, is not a just a comparison between then and now. This book has an additional purpose, of bringing out the incredible deduction process that allowed Aristotle to make inferences about scientific ideas based on a few fundamental assumptions, on non-instrumental observations and on logic. In writing this book, we consulted two translations from Ancient Greek: one into English (*The Complete Works of Aristotle: The revised Oxford Translation, edited by* Janathan Barnes, Bollingen Series LXXI 2, Princeton University Press 1984) and one into modern Greek (*The complete Works of Aristotle,* Volumes 13 and 14*, edited by* H. P. Nikoloudis, Cactus Editions 1994). For our discussion of rainbows in Book C, we found the paper 'The Aristotelian Explanation of the Rainbow' by A. M. Sayili, *Isis* Vol. 30, No. 1: 65-83, 1939, to be a very informative reference. We note here that during the last 2,300 years and after two translations (to Arabic and Latin), some of the style and concepts have been naturally affected.

That is why we combined the English and Greek translations and our familiarity and knowledge of ancient and modern Greek and meteorology to present accurately the writings of Aristotle. Our version of Aristotle's writing may at times appear 'exotic', partly because the style of expressing in Ancient Greek (like any other ancient language) was very different to the style used today, much like today's English differs from Shakespearean English. In Greek language, for example, it is very common (even today) that several tenses will be used in one sentence. Nevertheless, we think that our version has a flavour to it, and we hope the reader will adapt it without much effort.

Anastasios Tsonis and Christos Zerefos
Wisconsin and Athens 2020

Introduction: about Aristotle

His life

Aristotle, son of Nicomachus, was born in 384 BC in Stagira of Macedonia, about 55 km east of today's Thessaloniki. He died in Chalcis, a city on the island Euboea (Evia) just west of Attica, in 322 BC. His father was the doctor to Amyntas II, the king of Macedonia, who was the father of Phillip II and grandfather of Alexander the Great. His mother, Phaistis, descended from Chalcis. Both his father and mother belonged to the family of Asclepius. Aristotle's parents died when he was about thirteen and Proxenus of Atarneus became his guardian. Very little is known about Aristotle's early life. It is not even certain how much time he spent within the Macedonian Palace. At the age of 17, Proxenus sent him to Athens to study at the Academy, the school established by Plato. There Aristotle remained for 20 years. In 348 BC Plato died and was succeeded by the son of his sister Potonis Speusippus. For this reason, as well because the relations between the Athenians and Phillip II had worsen, Aristotle and his friend Xenokrates, left Athens and settled in Assos (Asia Minor), south of the Trojan coast across the island of Lesbos. There Aristotle met Hermias of Atarneus whom he initiated to Platonian philosophy, and who became a very close friend. Aristotle stayed in Assos for three years (348-345 BC) and then, after an invitation by Theophrastus, moved across to Mitilini where he stayed until 342 BC. That year he was invited to take over the education of young Alexander, who later became Alexander the Great.

Aristotle remained in Macedonia until 335 BC. During this time, in addition to educating Alexander, Aristotle embarked on to theoretical studies, which were interrupted in 341 BC by the death of his friend Hermias. "In his honour Aristotle wrote his only surviving poem *Ode to Virtue* (Αρετα πολυμοχθε...).[1] After 335 BC, Alexander now being the King of Macedonia, Aristotle moved to Athens again, and established the Lyceum (or Peripatetic School). He stayed there teaching and writing until Alexander's death

---

[1] The poem is included in the Appendix I at the end of the book.

in 323 BC. Subsequently, his position in Athens became difficult (among other things he was accused of insulting the Gods). Aristotle was forced to leave and went to Chalcis where he died the next year at the age of 62. The Stagirites moved and buried his body in his birth place.

His works

Aristotle wrote about basically everything. The complete catalogueue of Aristotle's work, accepted today, is by August Immanuel Bekker of Berlin University, published in two volumes in 1831. Three more volumes were added at a later date. Bekker based his work on three ancient catalogues by Diogenes Laertius, Ptolemy Henou (most likely a student of the Peripatetic school), and Isychios. Not all the works by Aristotle are listed in this book; Aristotle wrote tens of titles, which account for about half a million lines.

In his books Aristotle covered all philosophical topics and questions: nature, being, virtue and ethics. These subjects were studied in combination with the critical research of the political evolution of Greek city-states. However, Aristotle did not act only as a philosopher, he went a long way from supernatural explanations and beyond superstition. He was a master of inductive, systematic, and practical thinking, and was unparalleled in his ability to organise events and evidence. He contributed immensely to Plato's ideas and was determinative in the development, evolution, and application of the philosophical systems, that have influenced and still influence the world today. For Aristotle, there is no 'ideal' society, but only 'noble' (best possible) society within human capabilities. He considered education and virtuous upbringing as the best means for survival and progress.

The works of Aristotle were transferred from the Library of Alexandria to the Library of Antioch after the fall of Alexandria to the Arabs in AD641. After the conquest of Antioch most of the Alexandrian philosophers were converted to Islam and a large number of them immigrated to Spain. As early as the 9th century AD, these works were translated first into Arabic and later into Latin. They spread all over Europe. Aristotle's works were not studied only by philosophers but also by renowned scientists such as Galileo, Newton, Keppler, among many, who took counsel from his works and methods. Aristotle's views were adapted both by Christianity and Islam, as seen by miniatures in ancient theological manuscripts from major Christian and Islamic sources, such as *Sinaitic Codices* and *Islamic Codices* at Azhar Cairo (the first Islamic University in the world). Aristotle's views formed the basis for the study of physics and natural sciences for more than 1500 years and were adapted by both Christianity and Islam. Since the 19th century, no education could be considered complete without a reference to this great man from Stagira, the father of philosophy, and one of the greatest (if not the greatest) minds of all time.

BOOK A FROM ΜΕΤΕΩΡΟΛΟΓΙΚΑ

Aristotle's universe with a glimpse on climate change

A good starting point to Aristotle's writings is his view of the Universe. This view is fundamental not only in his discussions in *Meteorologica* but in other topics he wrote about as well.

It may not be an exaggeration to say that ever since humans were able to put one plus one together, they have been fascinated with the 'heavens'. The sun, the stars, the moon and the eclipses raised human curiosity to the maximum. All these celestial objects and phenomena needed an explanation. Thus these objects were perceived as representations of the heavens where the gods reigned. The need to know the gods better was established.

While many ancient cultures such as the Sumerians and Babylonians observed and recorded the movement of celestial objects, it was the Greeks in the mid first millennium BC who really advanced astronomy. In order to explain the effects of the horizon, Anaximander, in as early as the 6th century BC, suggested that the planet was round. Both Plato and Aristotle advanced the idea of a spherical earth. Aristotle's model of the universe was that each of the planets, the Sun, and the moon were set on a crystalline revolving sphere (Figure 1). The spheres were concentric with Earth at the centre. The stars were placed in the sphere, that surrounded all other spheres. The outermost sphere was, according to Aristotle, the domain of the 'Prime Mover'. The Prime Mover propelled this outermost sphere into a rotation that was passed on from sphere to sphere causing the whole universe to rotate.

Given this view of the universe, we are ready to begin our exposition on Aristotle's incredible take on meteorology. We will start with Book A, which is divided into 14 chapters.

ARISTOTLE'S Μετεωρολογικά

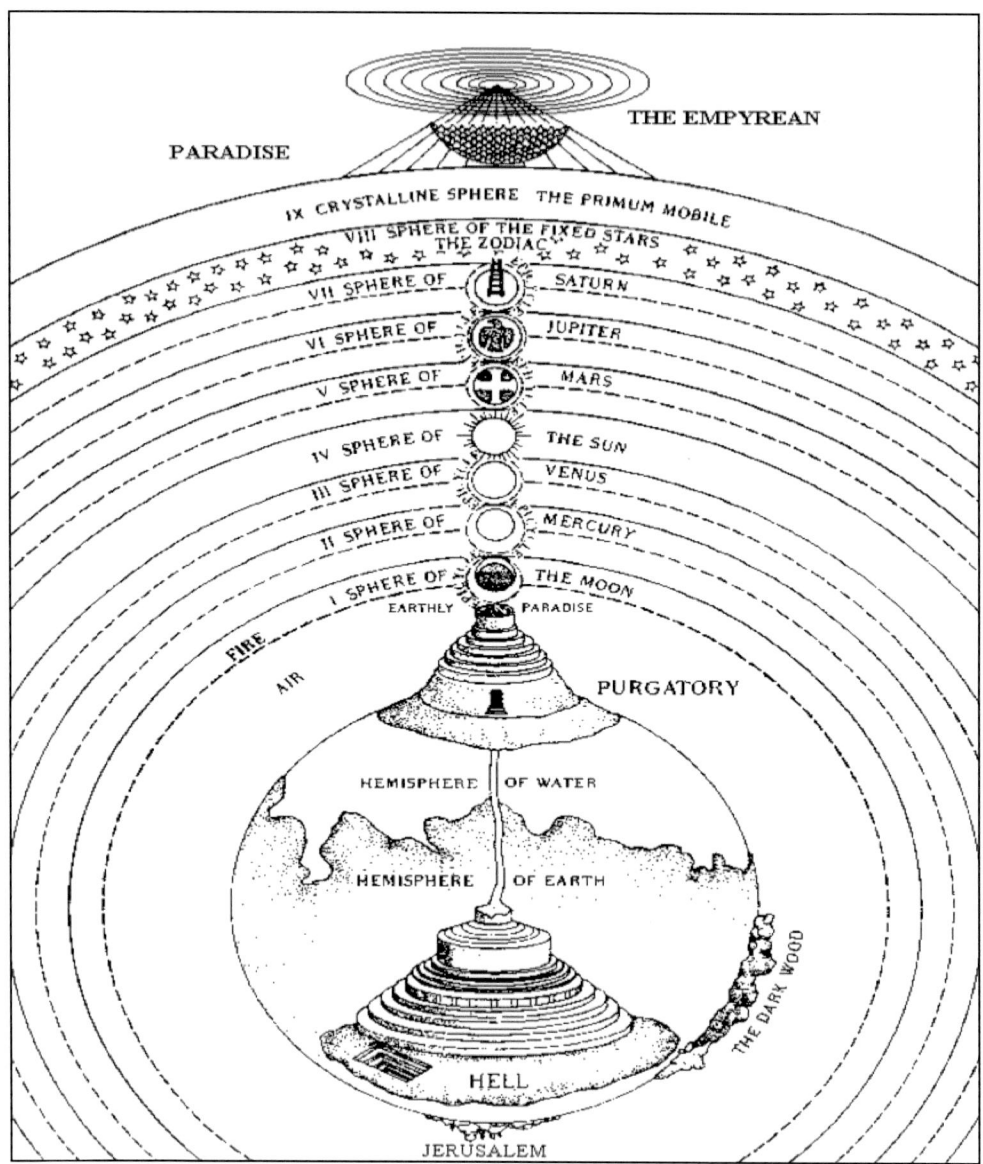

Figure 1: Aristotle's view of the universe lasted for 2,000 years and had far reaching consequences. From the edition by Thomas Digges of his father's *A Prognostication everlasting...*, published in 1576 in London (by permission of the Royal Society). The solid lines define the spheres and the broken lines their motion (orbit).

Chapter 1 is just an introduction where according to Aristotle, meteorology is concerned 'with all phenomena that obey natural laws, though the order they are subjected to is not as strict as that of the first element, and they take place in the region closest to the region where the stars move…'. The 'first element' refers to *aether* (or *ether*). We will come back to this soon.

In its entirety chapter 2 reads as follows:

> 'We have clarified elsewhere,[1] that there is an element associated with the bodies that move in a circle[2] and four principles,[3] which produce four additional elements whose motion is double, either from the centre (centrifugal) or to the centre (centripetal). These four elements are *fire, air, water, and earth*.[4] Among them, fire occupies the highest place and earth the lowest. The other two relate to fire and earth as follows: air is always found closer to fire and water closer to earth. These are the elements from which the world surrounding the earth is made of. It is the changes of this world that we must study. This world is necessarily, in one way or another, connected to the motions of the upper world so that all its changes are dictated from these motions, whose originating principle must be considered as the first cause. This cause is eternal and its motion is not restricted in space. It is perfect, whereas all other objects have restricted space which limits one another. Consequently, we should consider fire, earth and any other element like them, as the material cause of the phenomena that influence this world (we call 'material' what is subject and affected), but the originating cause of their motion should be searched for in the impulse provided by the eternally moving bodies'.

If this translation still sounds like Ancient Greek you are not alone! Some clarifying notes are in order.

Aether comes from the Ancient Greek αἰθήρ, which means 'pure fresh air'. It is also called *quintessence*, and it is the material which, according to Aristotelian view, fills the region of the universe above the terrestrial sphere (which includes earth, water and air), where the heavenly bodies move circularly (see Figure 1). Note that, in translations and in other works in Aristotle's writings, aether is often referred to as the fifth element after fire, air, water, and earth. Here the aether is referred to as the first element.

---

[1] In his other books.
[2] Aether.
[3] The four principles are: hot, cold, dry, and moist
[4] This notion was first introduced by Empedocles (5th century BC)

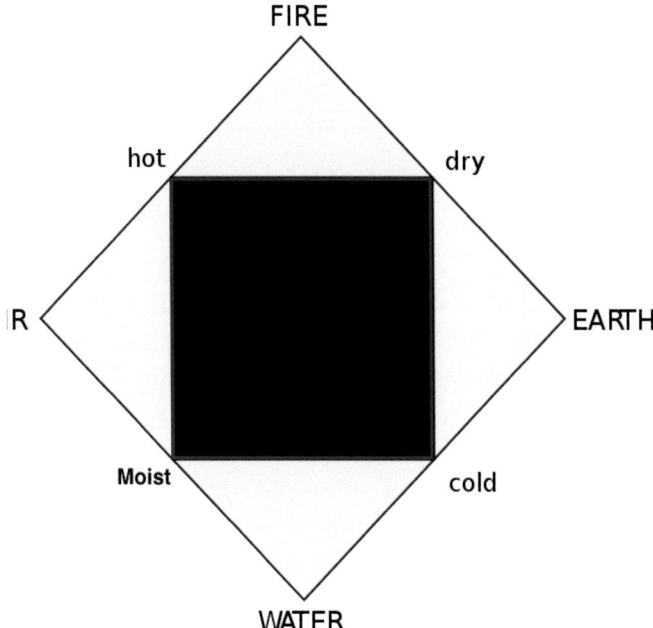

Figure 2: The relationship between the principles and the four elements. The corners of the inner square correspond to the four principles and the corners of the surrounding square to the four elements.

The four principles are *hot, cold, dry,* and *moist*. They combine to create in the sub-lunar world four elements, *fire, air, water,* and *earth*. Figure 2 illustrates the relationship between the principles and the four elements. The corners of the inner square correspond to the four principles and the corners of the surrounding square to the four elements. Accordingly, Earth is the result of the combination of dry and cold, air of hot and moist, fire of hot and dry, and water of moist and cold. While not discussed here (it is elaborated in Book D), Aristotle argues that hot and cold determine, conjoin, and change things. Thus, he considers them as active, for combining hot and cold is a sort of activity. Things dry and moist, on the other hand, are the results of that determination. Because dry and moistare being acted upon, they are passive. The four elements make up the sub-lunar world. Note, that the element of 'fire' does not refer to flames. Rather, it refers to an abstract entity resulting from the hot and dry principles.

The sub-lunar world is a continuation of the region above where the heavenly bodies exist and rotate.[5] No space exists between them, therefore because they 'touch' each other, the sub-lunar region is directly affected by the region above it. The circular

---

[5] It is interesting to note that as mentioned in Plato's *Phaedo*, Socrates conversing with Simmias at his very last moments, just before he drunk the conium, said that this region would be the region were his soul will continue to exist looking down at the spherical earth.

motion of the heavenly bodies is unrestricted by space since the objects always return to the same point, *ad infinitum*. Thus, it is perfect.

Chapter 3 is rather complex. For this reason, we will not present a translation (we found both the English and the Greek translations rather tedious). Instead, we will present what Aristotle is trying to convey in layman's terms. In a sense, in this chapter Aristotle is trying to derive the distribution of the elements in his universe.

Aristotle starts by stating that the four elements, fire, air, water, and earth, come-to-be from each other, and each of them can exist in each. His underlying assumption here is that there is some kind of balance between these elements. He then states that water is never seen collectively isolated nor can be found away from the surface of Earth, where is found as seas, lakes, or rivers. He then poses the first question: is the intermediate region between Earth and water and the outer sphere (above the moon) filled of one unique in its nature element or of more? To answer this question, Aristotle begins by referring to current understanding of the period that the region where the heavenly bodies move is filled with aether. Here he refers to Anaxagoras who identified, a century before Aristotle, aether with fire or as Aristotle calls it 'source of power' or 'stock of strength'.[6] Here Aristotle is not clear in his differentiation between aether and fire. Be this as it may, Aristotle argues that the space between Earth/water and the outer sphere cannot be filled with this power, because if that was case, all elements will have seized to exist (obliterated) long ago. He then argues that this space cannot be filled only with air either, because even if there were only two elements to fill the intermediate space, air will completely dominate and that would tip the balance, which is based on the proper proportions to the other elements. Thus, he concludes, neither air nor fire alone fills the intermediate space. Both have to be present.

Having established this, Aristotle then poses further questions: what is air's nature in the world surrounding Earth? What is its place in relation to other elements? How are air and water distributed in relation to the rest of the elements?

He begins his elaboration on the above questions by asking another question: since water and air are interchangeable and one is produced from the other, how come we do not observe clouds in the upper atmosphere? This is something that should be happening, he argues, since the upper atmosphere is colder, being relatively far from both the heat from the stars and the rays reflected from the earth, which rays, furthermore, should not allow clouds to form close to earth's surface as their heat will dissolve them. Rather, he asserts, clouds form at an intermediate level between

---

[6] Aether is related to αἴθω which means 'to incinerate', and intransitive 'to burn, to shine' (related is the name Aithiopes (Ethiopians) meaning 'people with a burnt (dark) face').

the surface and the upper atmosphere by which level the rays from the surface have been dispersed and thus are not effective anymore. Given the above, Aristotle ends up with two options. Either not all air produces water and clouds, or what surrounds the Earth has uniform properties and it is not just air but some kind of vapour which condenses to water. In other words, there are two possibilities: 1) there are two layers of air, one in the lower atmosphere and the other in the upper atmosphere. In the lower layer vapour in the air condenses and produces clouds, but in the upper layer it does not because it is dry; 2) all air can condense but the layer close to the earth has higher amounts of vapour, hence is more likely to produce clouds. Thus, what must immediately surround Earth is air and vapour.

At the same time, he continues, the circular motion of the aether-filled upper region of the universe generates heat, which inflames the lower world. The lower world is made up of some matter which obeys the four principles (hot, cold, dry, moist), and has all the properties that derive from them. However, it only retains the properties which depend on the motion of the upper world or absence of motion in the lower world. As a result of the absence of motion in the lower world, the heavier and colder elements (earth and water) sink to the centre with the lighter elements (air and fire) rising above them.[7] Air close to the Earth is warm and moist (owing to exhalation of water; exhalation is properly defined at the beginning of chapter 4). At higher levels on the atmosphere air is warm (from the heat by the stars) and dry (from being far from water). Alternatively, Aristotle adds, nothing is stopping us from arguing that it is the rotation of the heavenly bodies that inhibits cloud formation in the upper atmosphere. This rotation, by continuity, affects the upper levels of the atmosphere, thereby supplying them with heat. Either way, the upper atmosphere is either mostly heat or too dry to produce clouds.

Having dealt with the above question, Aristotle then asks: how does the heat generated by the stars in the upper world reach Earth and its neighbourhood? Heat is, as mentioned above, generated because the heavenly bodies are in motion. And the faster the motion the more the heat generated. From all these bodies, the Sun is the most dominant in warming the Earth. The rest are either too far or their motion is too slow. That is why the moon, while the closest to earth, does not contribute to Earth's warming; its motion is too slow. Subsequently, the heat from the Sun disperses, due to its rotation, and then is funneled toward the Earth.

This is how Aristotle's logic arranged the distribution of elements in his model of the Universe. A simple illustration is provided in Figure 3.

---

[7] First direct mention that warmer elements are lighter, colder are heavier.

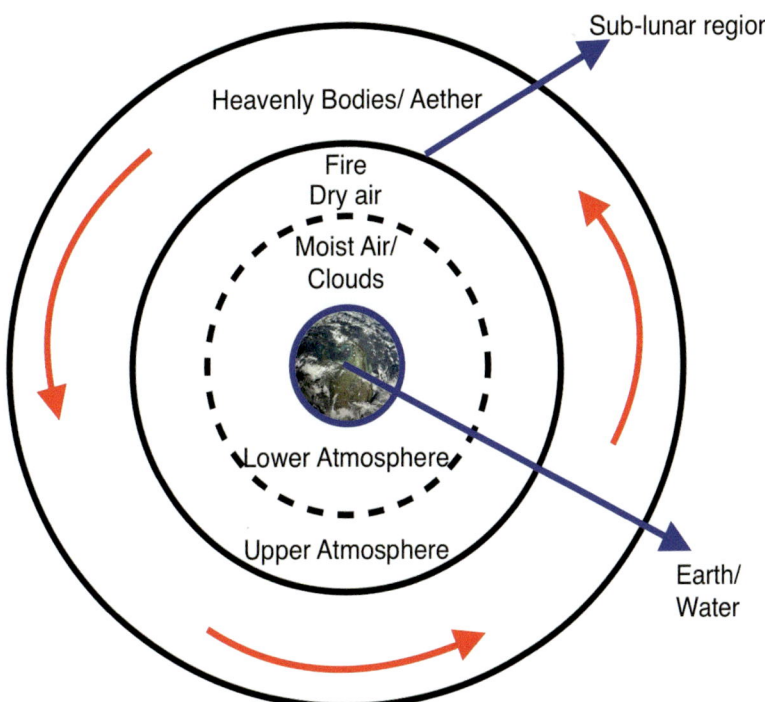

Figure 3: A simplified version of Aristotle's universe and the distribution of the five elements (aether, fire, air, water, and earth). The red arrows indicate the rotation of the upper world (heavenly bodies).

In Chapter 4 Aristotle defines a very important and paramount to his writings (briefly mentioned in chapter 3) phenomenon of *exhalation*. 'When the Earth is warmed by the Sun, causes it to exhale or "breath out". This exhalation is not of one kind, as many think, but of two. One kind manifests itself as vapour and the other as air. The moisture springs out of the water covering the Earth and of moisture inside the land (moist exhalation). The other kind is generated from the dry parts of earth (dry exhalation). From these two kinds only the second is able to rise because it is warmer[8] whereas the first one being moist stays on the surface because of its weight.'

This postulate provides Aristotle with additional 'evidence' for his distribution of elements in the sub-lunar region. He continues: 'It is for this reason that the world surrounding the Earth displays the order we discussed above. Exactly under the circular motion of stars, the warm and dry element (which we call fire) is found.... Under that we find dry air and below dry air we find moist air.'

---

[8] First direct mention that warmer air rises.

Aristotle closes this chapter by elaborating that the fire in the upper atmosphere is some kind of very flammable entity and that the smallest disturbance will cause it to burst into flames. He uses this belief to argue that this is how phenomena such as 'torches', 'goats', and 'shooting stars' appear in the upper atmosphere. These phenomena may also be observed at lower levels, but in this case their cause is not because of the fire but because of the release of heat due to condensation of vapour.[9] He concludes that the material cause of these phenomena is exhalation, whereas the effective (the mover) cause is sometimes the upper motion and sometimes the condensation. In the following 4 chapters (5-8) Aristotle employs his ideas based on exhalation, fire, and their effects, to elaborate on the causes of other astronomical bodies such as comets and the Milky Way. As we mentioned in the introduction these topics have little to do with meteorology and we will omit discussing them, with the exception of the phenomenon of Aurora Borealis, which we will discuss at the end of Book C in which Aristotle discusses other light phenomena like halos, rainbows etc.

Chapters 9-12 deal with precipitation. We will treat precipitation in detail separately, but at this point we will discuss chapter 9, as it touches upon some concepts that relate to chapter 4. In Chapter 9 Aristotle states: 'Paramount to our discussion is the cyclic motion of the Sun.' Clearly Aristotle here refers to the daily cycle of the Sun. 'Since Earth is not moving, water on its surface evaporates, due mostly to the heat from the Sun, and it rises. When, however, this heat is gone, because it disperses at higher levels or because it is simply quenched by the air, its vapour in the rising air condenses[10] and becomes liquid water and falls back to Earth.[11] The exhalation from liquid water produces water vapour and its condensation makes a cloud. Fog is what may remain from the condensation into cloud. This is why fog is likely a sign of fine weather rather than of rain. Fog is something like a barren cloud. These changes in water phases follow the cyclic daily motion of the Sun. Truly, depending on how high or low the Sun is in the sky, the amount of vapour fluctuates. We must think of such a phenomenon as a river moving up and down in a circle. When the Sun is closer (overhead), the stream of moist air flows upwards, and when the Sun is further (later in the day), it flows down. And this phenomenon is reproduced in the same order without exception. Therefore, if earlier writers were attaching some secret meaning to 'Oceanus' (ocean), they probably meant this river which flows in a circle around the Earth, 12.13 To summarise: moisture rises because it gets warm and falls back to

---
[9] First mention that condensation is associated with heat release.
[10] First mention that as air cools its vapour condenses.
[11] Definition of the hydrological cycle.
[12] This refers to Herodotus (484-425 BC) who wrote 'The Greeks speak of "Oceanus" that it stems from where the Sun rises and flows around the Earth, but they don't prove it.' Note that the mention of Oceanus also appears in Plato's *Phaedo*, in a discussion between Socrates and Simmias.
[13] First notion of what we call today convection, a circular motion of warm and cold air, which is here suggested to occur in small (local) and in planetary scales.

Earth because it gets cold. Special names have been given to these processes and their varieties; falling small drops are called drizzle and bigger drops are called rain.'

In summary, we may list as the major inferences in chapters 1- 4, 9, the following:

1. The universe is geocentric with all other 'heavenly bodies' revolving around Earth.
2. The distribution of the four elements and aether in Aristotle's universe shown in Figure 3.
3. The Sun warms the planet and water evaporates.
4. Notion of moist and dry exhalation. Atmosphere as a two-component system.
5. Dry exhalation rises because it is warmer. Moist exhalation is heavier and sinks below the dry exhalation.
6. Warm air is lighter, cold air is heavier.
7. Warm air rises, cold air sinks.
8. When it is cold, water vapour condenses to produce water and clouds.
9. Condensation releases heat.
10. Notion of convection.
11. The hydrological cycle.

Let us consider the first inference. Aristotle's view of the universe with Earth at the centre (called the geocentric model; geo means Earth in Greek) is based on the belief that the Earth was not moving, a belief common for the period of writing. This belief was based on the available observations at the time. For example, the heavens were out there and Earth was not part of them. The heavens showed little change. The stars were at the same place every night. In contrast, Earth was always changing. They attributed this to a regularity of the heavens, which could never be observed in corruptible Earth. How can then Earth be part of the heavens? In addition, objects in the sky were luminous whereas Earth was not glowing. How can Earth be similar to the heavens? Finally, they observed that in the atmosphere clouds were not left behind, as they should if the planet were in motion. They also argued that when you jump you land on the same spot. How can this be so if the place is moving? Today we understand all these observations, but at the time they presented indisputable evidence that the Earth was motionless. Thus, Earth must be the centre of the universe with everything else moving around it in perfect circles.

Aristotle's view, even though incorrect, allowed him, by applying mathematics, and especially geometry, on what he saw of the stars' movements, to predict the movements of the planets. However, it soon became clear that there were problems with Aristotle's view. His view could not explain the so-called retrograde motion of the outer planets. This motion occurs when the Earth catches up and overtakes an outer planet. This phenomenon can be explained very easily if the Sun is at the centre and the planets revolve around it (heliocentric model). Because the orbit of Earth is smaller than that of the outer planets, as we go around the Sun the planet catches up with and passes the outer planets, which now appear as if they move backwards. Interestingly, a student of Plato, Aristarchus of Samos (often referred to as the Copernicus of antiquity), was the first to propose a heliocentric model of the universe in the early third century BC. His model, however, was dismissed at that time. Unfortunately, the great popularity of the geocentric model did not foster such new and revolutionary ideas. Rather, in order to explain these backward motions and to fit the observations, astronomers proposed complicated additions to the geocentric model. Ptolemy, who lived from AD 85 to AD 165, revised the model by assuming that, as a planet moves around the Earth, it is not only moving along a circle with Earth on its centre but that it revolves on a smaller circle (the epicycle) whose centre moves along the circle with Earth in its centre (Figure 4).

This model fitted the data at the time well, but as more data were gathered, more corrections and more epicycles had to be added, thereby complicating the picture. Nevertheless, the geocentric model lasted for nearly 2,000 years. Then, around the mid-second millennium, Polish astronomer Nicolaus Copernicus (1473-1543), reasoned that a complicated universe cannot represent the elegance and wisdom of God. Copernicus was a Neo-Platonist who believed that God not only was the creator but he was a wise creator who would have not made a complicated universe. He also believed that because the Sun provides warmth and light (necessary ingredients for life), it is a copy of God. Thus, the Sun must be at the centre of all. Motivated by his

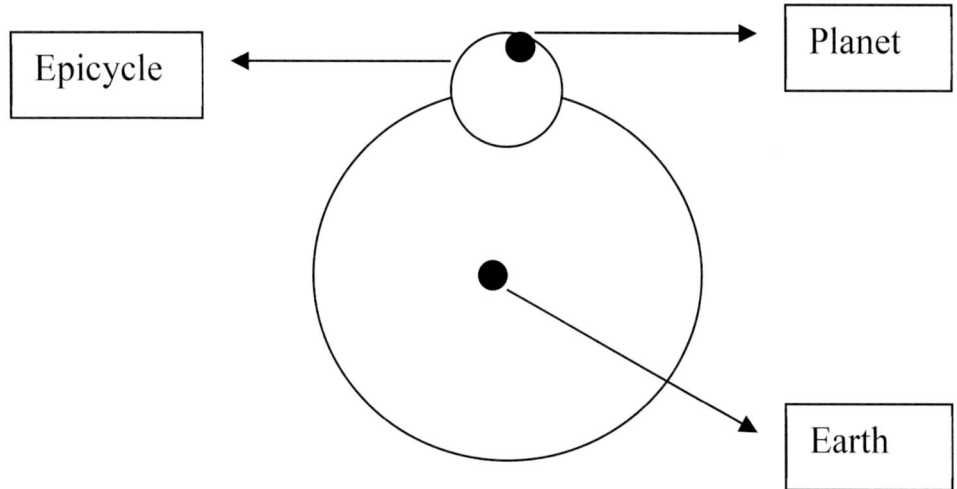

Figure 4: Ptolemy's model to explain the retrograde motion of planets.

belief, he adopted Aristarchus' heliocentric model. He also found that the farther the planet from the Sun the slower its motion, and he was thus able to explain the retrograde motion without the need of the annoying epicycles.

Copernicus' model, however, would not go well with the Church. At that time the Christian leaders were firm believers of the Aristotelian view of the universe with its imperfect Earth and perfect heavens. The Prime Mover had been identified with God and his crystalline sphere with the Christian Heaven. Moving away from the centre meant perfection, more control, somebody to watch over you. The Church's paradigm, that Man is God's special creation of the physical universe and Earth the centre of a wonderfully planned universe, was not compatible with Copernicus' model. Copernicus knew this and that is why he delayed publication of his model until 1543, the year of his death.

A few decades later came Galileo (1564-1642), with the help of the telescope, made new and improved observations which clearly demonstrated the validity of the heliocentric model. He also ran into troubles with the Church. He was summoned and to save his life he was forced to give up his work and recant his theories.

But the tides had changed. Once a theory is solidly demonstrated there is very little to argue about. Today, there is no educated person who thinks otherwise (except those who believe that the Earth is flat!!!).

Thus, Aristotle was wrong when it comes to his view of the universe.

That aether permeates space, was a common belief in the scientific community up to the end of the 19th century. It was needed to explain how light waves are carried in space and how gravity can act between distant bodies via propagation through an intervening medium. It was not until the Michelson–Morley experiment in April 1887, and the subsequent theory of relativity by Albert Einstein, that the proof that aether is not necessary for light to travel in space or for gravity to act between two bodies in an empty space was given. Soon after that aether became history.

The rest of the inference will be examined collectively after a little tutorial on what we understand today about the contents of Book A.

Meteorology now, part 1

Our atmosphere has not remained the same in time. Early in its history, the atmosphere consisted of hydrogen and helium, which were probably supplied by the solar wind carrying them from the sun to our planet. Later, when the planet developed a differentiated core (solid inner/liquid outer core), the earth's magnetic field was established, which deflects solar wind. This cut off the supply of hydrogen and helium. At that time, increased volcanic activity introduced many other gases such as water vapour, nitrogen, carbon dioxide, ammonia, methane, etc. Condensation of water vapour produced liquid water, which filled the oceans and photosynthesis (a chemical reaction between carbon dioxide and solar radiation) produced organic compounds and oxygen.

The various constituents of the atmosphere have different atomic weight. Because of the planet's gravity, all objects much smaller than the Earth will accelerate toward the Earth's centre of mass (toward the Earth's core). Because the gases in the atmosphere are not able to go through the Earth's crust, the majority of them are compressed near the Earth's surface. This is why the atmosphere is most dense in the low levels of the atmosphere and very thin at high levels (Figure 5). In addition, heavier elements are pulled more and are thus found closer to the surface. On the contrary, light elements are found at higher altitudes. This divides the atmosphere into two regions the lower region (below 50 miles) called the *homosphere* and the upper region called the *heterosphere*.

These two regions are then arranged into four layers. The lowest layer (the one closest to the surface) is called the *troposphere*. The word derives from the Greek words τρόπος and σφαίρα, meaning behaviour and sphere, respectively. This is the layer where weather manifests itself. It is the layer where all weather (rain, snow, wind, storms, tornadoes, hurricanes, etc.) takes place. It contains 75% of the atmosphere and 99% of water vapour. No weather is observed above the troposphere. Next layer is the stratosphere. Stratosphere derives from the words *stratum* and *sphere*. Stratum in Latin means uniform. The name very clearly characterises this layer as it is a layer where there is hardly any movement of air. This is where airplanes prefer to fly, so that they are not exposed to headwind. In the stratosphere it is very dry (too far from the source of water vapour, which is water evaporating from the surface), and it is there where we find the important gas ozone. Above the stratosphere is the *mesosphere* (from the Greek word μεσαίο meaning middle). The upper and final layer is called the *thermosphere*. As the name suggests, thermosphere is a very hot place. This is because this layer, being the outer layer, intercepts the complete (unaffected from scattering and/or absorption) radiation. It thus has more energy available to warm up. Since the air is very thin there, the total mass is small and even small amounts of radiation absorbed result in high temperatures. In addition, this is the area where the very energetic x-rays and gamma rays are being scattered. Their high energy strips the atoms of their electrons thereby producing a lot of ions (positively charged atoms and negatively charged electrons; in the past this part of the atmosphere was called the *ionosphere*). This process is associated with heat release, which also helps warming the thermosphere.

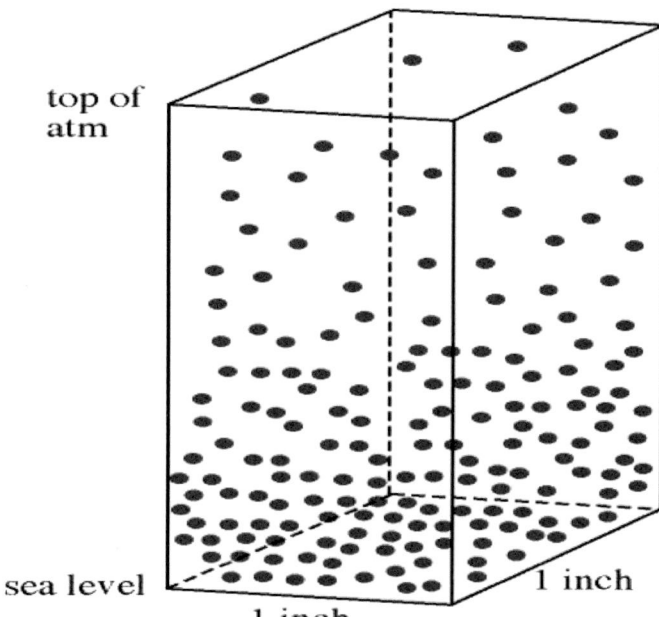

Figure 5: Distribution of particles with height. Because of gravity the atmosphere is denser close to the surface. As we go higher density decreases.

In the homosphere, our atmosphere consists mostly of nitrogen and oxygen. Seventy eight percent of the atmosphere is nitrogen and 21% percent is oxygen. All other gases make the remaining 1% with argon claiming 93% of that one percent (i.e. 0.93% of the total atmosphere). Another gas that is found in the atmosphere is water vapour. Water vapour is not a permanent but a variable gas. It is introduced in the air through evaporation of water when water is warmed up by solar radiation and it is removed via precipitation. We all appreciate the importance of oxygen. Apart for keeping us alive, it is a major constituent of water, of most rocks and minerals, and of the human body. Nitrogen is as important, since it helps dilute oxygen in burning and respiration processes. Also, bacteria in the soil use nitrogen to produce nitrates, which are then absorbed by plants. What, however, we may not appreciate as much is that in that remaining one percent are included two gases without which life may not have developed on this planet. They are (not in order of importance) carbon dioxide and ozone (a relative of oxygen). Carbon dioxide makes up about 4% of the last one percent (0.04% of the total atmosphere). Ozone is found in even smaller amounts. It makes only 0.000004% of the total atmosphere. Both these gases play a very vital role in the atmosphere. Carbon dioxide acts as a blanket keeping the surface of the planet warm, and ozone absorbs the ultraviolet radiation which is harmful to life.

As we discussed earlier, because of gravity, the air density (the amount of mass per unit volume) decreases as we go higher. Individually, each molecule or particle in the atmosphere is very light but all together result in a considerable weight. Unbelievable, the total weight of the atmosphere is 5600 trillion tons. Assuming that the area of the top of one's head measures about 35 square inches, then the weight of the atmosphere in a column having as base the top the head and extending from sea level to the top of the atmosphere is about 500 pounds.[14] Thus, it is not only true that the air has weight, but it is actually quite heavy. This weight acts as a force upon the earth. When this force is divided by the area, we get the pressure that the atmosphere exerts on that area. This pressure is called *atmospheric pressure*. It is one of the most important variables in weather. As one climbs a mountain there is less and less atmosphere on the top of the head, therefore the atmospheric pressure decreases with altitude. This means that at a fixed level in the horizontal pressure is the same everywhere and also that the pressure at the surface of the planet is much higher than the pressure near the top of the atmosphere (where it is zero). Does this create a problem? Imagine pushing on one end of a table whereas a much bigger fellow is pushing from the opposite end. Which way will the table move? The answer is clear. The table will move toward the lighter person. The bigger fellow, because he is bigger, exerts more pressure. Therefore, on one end we have higher pressure and on the other end we have lower pressure. The net result: movement goes from high to low pressure. Every time a high and a low pressure are established, there is movement. In the atmosphere this movement is *wind*. The difference in pressure creates a force called the *pressure gradient force*, which provides the initial impulse that starts the air movement. Obviously, the higher the pressure gradient, the stronger the wind. Now, straightforward reasoning will tell us that the atmosphere should rush from the high pressure at the surface off into space. Of course, we know that this is not case. The reason is that this particular force in the atmosphere (the vertical pressure gradient force) is balanced by the force of gravity. This balance is called *hydrostatic balance*. Over large scales this balance is always maintained and the atmosphere stays where it is. However, as we will see later, locally (in small scales) vertical motions are allowed.

The atmospheric pressure at a given level (for example, on the surface) is not necessarily always the same. Imagine two identical rooms next to each other, each one equipped with two identical pieces of furniture. Because the pieces of furniture are identical, they weigh the same and exert the same pressure (P) on the two floors (Figure 6, top). If we take one piece from room 2 and move it to room 1, then the pressure on the floor in room 1 will increase (P1>P), whereas the pressure on the floor in room 2 will drop (P2<P; Figure 6, bottom). Similarly, in the atmosphere, if air is transported from some other area over to your area, your pressure will increase. As we will see later, changes in pressure are always associated with changes in weather. Incidentally, while changes in pressure in the vertical will create air movement in

---

[14] This fact was never realised by Aristotle. More on this later.

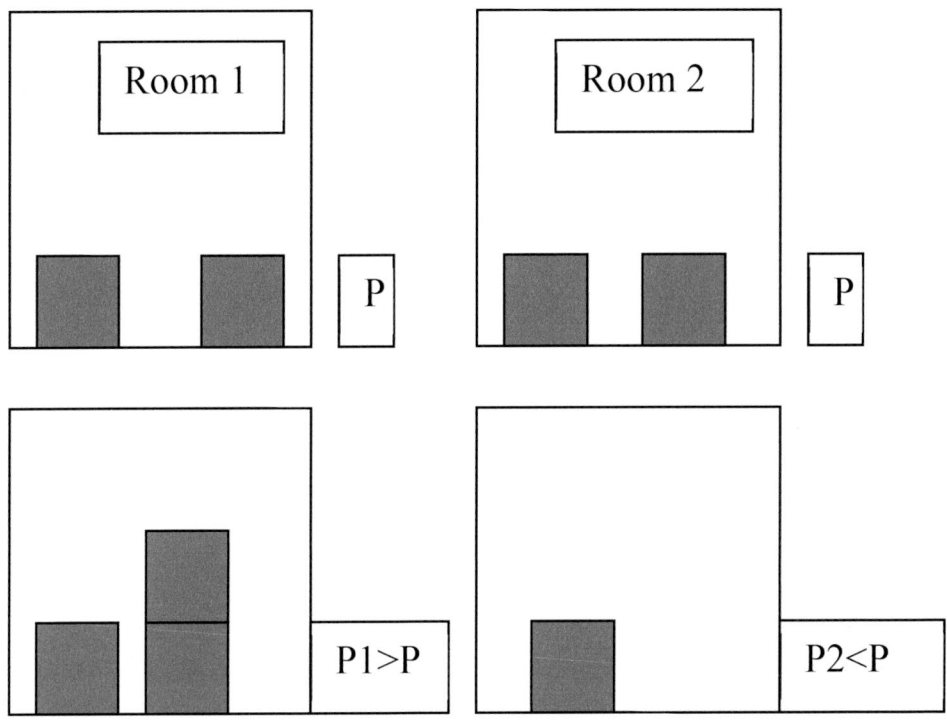

Figure 6: A simple demonstration on the relation of pressure and density.

the vertical and changes of pressure in the horizontal will generate movement in the horizontal, it is only the horizontal movement that is called wind.

It follows, that density and pressure must be related. Density is defined as the amount of mass per unit volume. After moving the piece of furniture to room 1, there is more mass in that room. Since the size of the room (volume) is fixed, this means that the density in room 2 increases as well. Given their definition, both density and pressure *decrease* as we climb the atmosphere.

Another variable related to density and pressure is temperature. We take temperature for granted, but exactly what is temperature? Defining temperature can be very frustrating. Searching the web for definitions of temperature makes one lost soon. Qualitatively speaking, we can describe the temperature of an object as that which determines the sensation we feel when we touch it.[15] Technically speaking, however, the definition of temperature varies. An object will absorb energy and become warmer, but exactly what happens to the object when the temperature increases? To answer this question, we have to see what happens to the molecules inside the object after they receive energy. Because of the extra energy, the molecules get excited

---

[15] Interestingly, in chapter 3 Aristotle considers temperature 'a certain affection of the senses'.

and move faster. This speed is what defines temperature. The faster the molecules move the higher the temperature of the object. Similarly, if the molecules slow down (for example, by losing heat), the temperature drops. As far as this book goes, this definition will be adequate. Based on this definition we can explain how temperature relates to density and pressure. If we imagine that the molecules occupy some 'box' before they are heated, then this volume will expand (like the air in a balloon expands when is heated). Since now we have the same number of molecules inside a larger volume, the density will decrease. What about pressure? Will it increase or decrease? The answer to this question can be understood if we, for a moment, assume that the 'box' is solid (like a closed room) and that the heated air is not able to push the walls aside. In this case, the density remains the same even though the temperature increases. However, the molecules are now moving faster and as a result, the chance to collide with the walls and floor increases. This 'banging' on the walls, increases the pressure exerted by the molecules. Thus, just because the density remains the same it does not mean that the pressure will remain the same. This would be the case only if the temperature remained constant. Thus, changes in pressure can come about from changes in density and from changes in temperature. The relationship between pressure, density and temperature is expressed by the *ideal gas law*, which states that the pressure of a gas is proportional to density and proportional to temperature. More specifically:

pressure = temperature x density x a constant

This will indicate that if density changes in some way while temperature is held constant, the pressure changes in the same way. Similarly, if temperature changes in some way, while density is held constant, pressure will change in the same way. If both temperature and density increase (decrease), then pressure increases (decreases). However, if density increases and temperature decreases (or vice-versa), then the effect on pressure will be determined by the one which changes more. The ideal gas law was derived in the 1800s by combining experiments performed by the French scientist Joseph Louis Gay-Lussac (1778-1850) and observations made by the English scientist Robert Boyle (1627-1691) almost a century earlier. It is one of the most fundamental laws in meteorology. The ideal gas law also implies that if the pressure remains constant while the temperature increases, then, in order to maintain the equality between the right- and left-hand side of the equation, density must decrease. It follows that when the air warms at a constant pressure level (for example surface), it must become lighter (because density must decrease). *Thus, air that gets warmer than its surroundings is lighter and as a consequence it will rise.*

Is there anything else that will make the air even lighter?

To answer this question, we have to bring water into the play. Seventy percent of our planet's surface is covered by water. Water is found in three phases: vapour, liquid and solid. At the surface of the planet the state in which water is found depends on temperature. When the temperature is relatively low, water exists as ice because,the molecules move at very low speeds and they cannot overcome the chemical bonds from their neighbouring particles. As a result, the molecules are held together in a rigid pattern. As the temperature increases the molecules move faster. Now, they are not bound to the neighbouring molecules so rigidly, but they are still bound strongly enough not to be able to escape. This way water becomes liquid. At relatively high temperatures all the bonds are broken and the molecules are free to move as they like, thereby becoming a gas. This free motion allows gas to expand and fill any space provided. As we all know, if we wish to melt ice or to evaporate liquid water, we have to provide ice or liquid water with heat. When these changes take place on a cook top, the burners provide the heat. When they take place in the atmosphere, the heat is borrowed from the environment. Therefore, melting or evaporation cools the environment. They are, in other words, *cooling processes*. That is why when one comes out from the sea or the shower one feels cold; water on the skin evaporates using heat from the body. Nature is very beautiful and wise, but also unforgiving. Thus, if it lends some energy for a change of phase (say, liquid to vapour), then when the opposite happens (i.e. vapour to liquid) this energy must be given back. It follows, that condensation and freezing release heat into the environment; in other words, they are warming processes. These 'hidden' heats are called *latent* heats of evaporation, of melting and so on and are very important in weather and climate.

Now, what happens when water vapour is introduced into dry air? A common misconception is that now the air has more 'stuff' in it and thus it should be heavier. This, however, is not true at all. In fact, the air is going to become lighter.

Back in the early 1800s, Lorenzo Romano Amedeo Carlo Avogadro (1776-1856), a lawyer turned scientist, discovered that a mole[16] of any gas, at a fixed pressure and temperature, and occupying some volume, will always have the same number of molecules regardless of the type of the gas. In other words, if we were to introduce some extra molecules in that fixed volume, then an equal number of molecules will have to leave the volume (of course this requires that the fixed volume is not rigid as, for example, a closed room but 'open' like a cubic foot of air in the atmosphere). This number is extremely large. To get an idea of how large it is, imagine the following.

---

[16] A mole is the amount of any substance which contains the same number of particles (atoms, molecules etc.) found in 12 grams of carbon-12. This number is obtained under the condition that one mole of carbon weighs 12 grams. In general, a substance of a weight equal to its molecular weight contains one mole of the substance. For example, given that the molecular weight of water is 18, then 27 grams of water contain 27/18=1.5 moles of water.

If a molecule were of the size of un-popped popcorn kernel, and the United States was to be covered with one mole worth of kernel, then the country would be covered in popcorn to a depth of nine miles. If one could count one million molecules per second, it would take more than the age of the universe to count all the molecules in one mole. When the air is dry, it mainly contains oxygen and nitrogen. Accordingly, when water vapour is evaporating in the atmosphere, some nitrogen and oxygen molecules have to be moved elsewhere. The molecular weight of dry air is determined by the molecular weight of nitrogen and oxygen. Nitrogen's molecular weight is 28 (two atoms with atomic weight of 14). That of oxygen's is 32 (two atoms with atomic weight of 16). Given that 78% of dry air is nitrogen and 21% is oxygen, it turns out that the molecular weight of dry air is about 29. The molecular weight of water is 18 (two atoms of hydrogen with atomic weight of 1 and one atom of oxygen). *Thus, water vapour is lighter than dry air, which means that when some dry air is replaced by water vapour to produce moist air, the molecular weight of air decreases.*

What we have learned up to this point is that warm dry air is lighter than cold dry air and that moist air is lighter than dry air. It follows that warm moist air will be much lighter than cold dry air. This bottom line is of paramount importance to weather. Baseball players who find it easier to score home runs in hot humid days than in cool dry days, know this fact very well. Less dense air provides less resistance for the ball, which means it travels farther.

Let us do a little experiment now. If you are close to the kitchen, pour some water in a pot and turn the heat on. While the water is warming up, cut small pieces of paper, crumble them and throw them in the pot. Wait until water begins to boil and observe what happens. The pieces of paper are caught in circular vertical motions that sink them to the bottom and then raise them to the top. These motions which are called *convection* appear to be everywhere. Every piece of paper is trapped in one of them. This simple process is of paramount importance to the atmosphere and weather. Let's see why and how it occurs in the atmosphere and what its consequences are.

The surface first warms up because of solar radiation. Then the air in contact with the surface warms up and rises. This way of transferring heat is called *conduction*. As warm air begins to rise, what are the changes occurring in the neighbourhood?

To construct the changes when warm air begins to rise, we need to recall the definition of pressure. At a given level the atmospheric pressure over a fixed area is given by the amount of the air included in a box having the fixed area as its base, and extending from the given level to the top of the atmosphere. Imagine two identical two-story houses adjacent to each other (top of Figure 7). On each floor of each house we place two objects of equal mass of one unit. If we ignore the weight of the walls and floors,

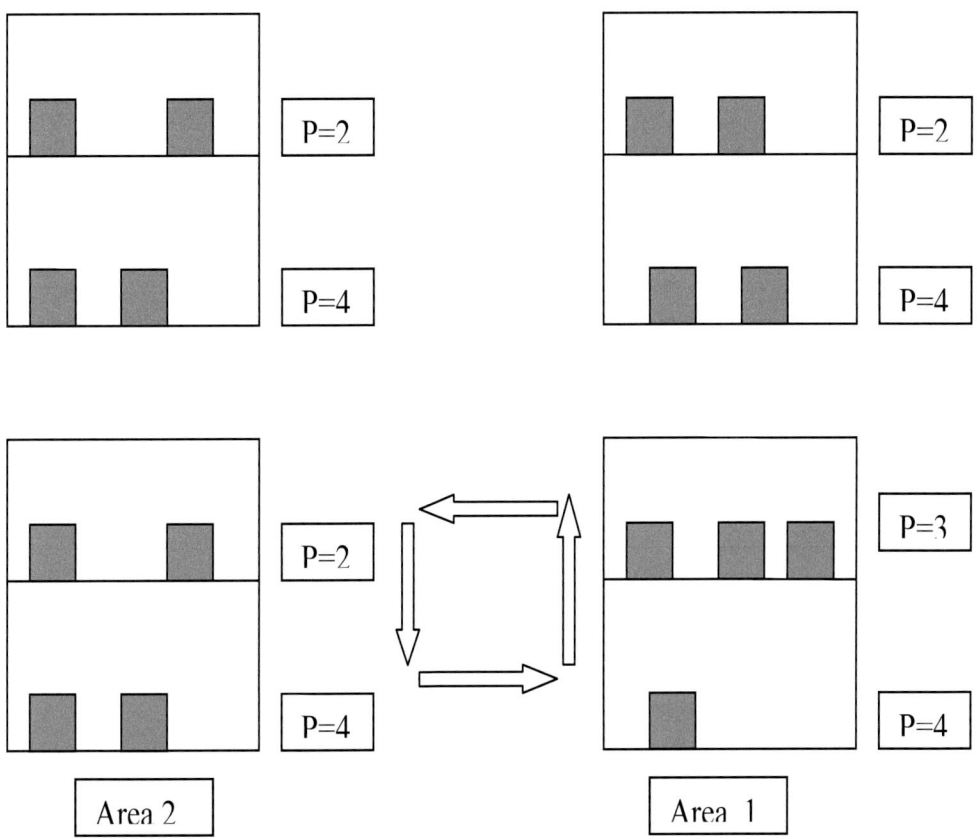

Figure 7: A simple example demonstrating how rising air will create convection.

the pressure at the bottom in each house will be 4 units and the pressure on the second floor in each house will be 2 units. In this arrangement pressure is *stratified*. It decreases with height and at any given level the pressure in the horizontal is the same. Now imagine that we take one object from the house on the right from the first floor and we move it in the second floor (bottom of Figure 7). Will the base of the house experience a change in pressure? Since the total pressure is the number of objects above the base, the pressure is still 4 units. However, the pressure on the second floor is not 2 units. It has increased to 3 units anymore. At the same time the pressure distribution in the house to the left has remained unchanged. In this case pressure still decreases with height but it is not stratified. At the level of the second floor, pressure is higher on the right house and lower on the left. Now replace the houses with two adjacent areas (area 1 and area 2) on Earth's surface and assume that initially the pressure is stratified. In other words, initially the pressure decreases with height but at any given level it is the same in the horizontal direction. Also, replace the unit objects with air. Then, from the above example it follows that when the air begins to rise over area 1, the pressure at the surface will not change, but the pressure at a higher level will increase compared to the pressure at the same level over adjacent area 2. So now, we are at a level above the surface and we have created an area of higher pressure and an area of lower pressure. What is going to happen next?

As we discussed previously, once a high and a low-pressure areas are established, a motion of air from the high pressure towards the low pressure ensues. In effect, this represents a removal of some air for area 1 and a deposit of air over area 2. This will cause the pressure at the *surface* in area 1 to drop and the pressure at the *surface* in area 2 to increase. This, in turn, causes the air at the surface to move from high to low to replace the rising air. At the same time the air sinks to replace the air removed at the surface over area 2, thereby closing a vertical circulation between the two areas (indicated with the arrows in Figure 7). Therefore, as the air over an area rises, it is soon replaced by colder air. The final result modification of the pressure at the surface; pressure drops where the air is rising and increases where the air is sinking. Thus, *rising air is associated with lower pressure and sinking air with higher pressure.*

Such circulations happen very often in the atmosphere. Consider for example what happens in the summer in Milwaukee or any other place close to a water body. During the day land warms faster than water and thus the air over land gets warmer compared to the air over Lake Michigan. So, the air begins to rise over Milwaukee. Soon after that cooler air from the lake flows to replace the warm air, thereby closing a circulation that we all know as *lake* or *sea breeze*. Later in the day, the situation reverses. As we begin to cool down, land cools faster than water and now the air will rise over the lake and this air will be replaced by a flow of air from the west. This circulation pattern is known as the *land breeze*. Similar circulations develop along mountain slopes where *valley* and *mountain breezes* are formed. The circulating pattern of warm and cold currents is called *atmospheric convection*, and is responsible for mixing and re-distribution of heat in the atmosphere. It is another way of transferring heat, together with radiation and conduction. Because temperature contrasts can occur at small as well as at large scales, these motions will take place at all scales even at hemispheric scales.

Because the planet is a sphere, the angle formed between the incident solar radiation and the tangent at any latitude is not constant. In general, equatorial regions intercept solar radiation more directly than polar regions. On the average the tropics receive more than the polar regions. This means that the atmosphere close to the surface will be warmer near the tropics and colder near the poles. This differential heating will cause a convection pattern of planetary proportions. The air around the equator will rise creating higher pressure aloft over the equatorial region, and lower pressure aloft over the polar regions. Thus, as with the lake or land breeze, the air in the upper levels will move toward the lower pressure regions, i.e. the poles. This removal of air

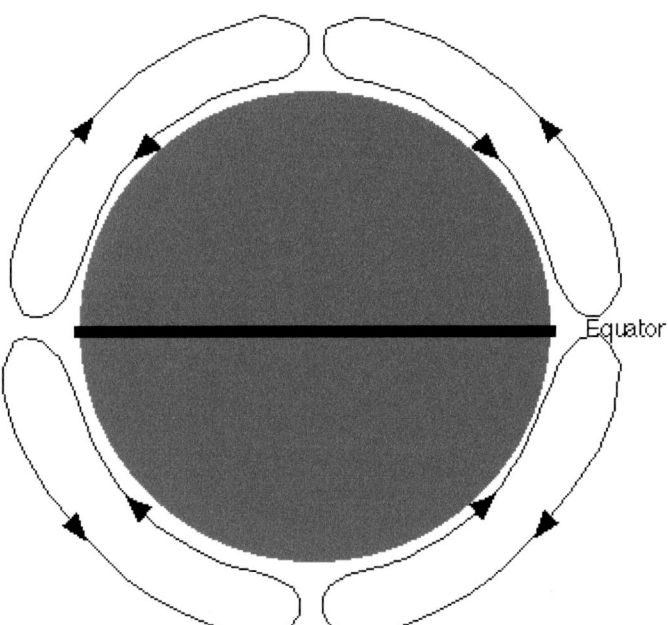

Figure 8: The Hadley circulation cells

from the equatorial region and deposition in the Polar Regions decreases the pressure at the surface at the equator and increases it at the surface at the poles. The result: a movement of surface cold air from the north and south toward the equator, and establishment of two simple circulation cells (one in each hemisphere) that will move the whole atmosphere (Figure 8).

This simple atmospheric circulation is called the Hadley circulation in honour of the eighteen-century English meteorologist George Hadley who proposed this idea. Embedded in each Hadley cell all smaller scale convection currents created by temperature contrasts at the earth's surface will be found. Our atmosphere can then be thought as little whirls within bigger whirls within even bigger whirls; one huge mass of circuitous elements of all sizes. Note, that the simple Hadley circulation, due to the rotation of the planet, is unstable and breaks down to three sub-cells in each hemisphere. We will not go into more details on that at this point. We will go into more details when we move to Book B and while discussing winds. For now, make a note that vertical movement of air initiated from convection, leads to changes in atmospheric pressure in the horizontal, which in turn leads to movement of air in the horizontal from higher pressure to lower pressure, which is what we call wind, and that the higher the difference between the high and low pressure the stronger the wind. And that, strictly vertical rising or sinking motion is not classified as wind.

A last topic to discuss, before we move on to the next chapter, is the vertical distribution of temperature in the atmosphere. How does temperature vary with height? If we follow what we have discussed so far and use our logic, we will decide that temperature must increase with height because as the solar radiation travels through the atmosphere it is been scattered and absorbed, which mean less and less radiation remains as it streams towards the surface of the planet. Compounding on this is the fact that closer to the surface the density is higher and therefore there is more mass to be warmed with less radiation available than at higher levels where there is less mass. Clearly then temperature must increase as we climb the atmosphere. Well, not really! This theoretical expectation does not happen. The temperature distribution of the atmosphere is rather complicated. It decreases with height in the troposphere, it increases in the stratosphere, it decreases in the mesosphere, and it increases in the thermosphere. Let' see why.

First let's consider what will happen to a *parcel of air*[17] when it begins to rise (Figure 9). To start with, we can imagine the cube of volume V1 sitting at the surface with its faces exposed to the pressure at the surface (P1). As the cube rises it goes into areas of lower pressure. This is because the pressure in the atmosphere decreases with altitude. For example, P2<P1. Thus, as the parcel rises the pressure of its surroundings lessens. Since now less pressure is squeezing the parcel, it is free to expand (V2>V1). Expansion means that the parcel is pushing away its surroundings, in the same way as the surface of an inflating balloon pushes the air around it. When we blow a balloon we do some work, isn't it? Well, similarly the parcel has to do some work during the expansion. Doing work means spending energy. Thus, the expansion of the parcel is followed by a reduction in its energy.

Now, what is the source of energy in the parcel?

Discussing energy can take us into some troubled waters, but let's give it a try. Our favourite definition of energy is *the ability to do damage*. Imagine a stone on a porcelain plate placed on the ground. The stone is doing no damage to the plate, which means it has no energy. If we raise the stone and then let it fall, it will break the plate into many pieces. Since the stone did damage, it must have had some energy. If we repeat this experiment but with lifting and letting the stone fall from a greater height, we can imagine that now the damage to the plate will be greater. This would indicate that the parcel had more energy when it is higher. Somehow this energy must relate to how far an object is from the surface of the planet. This energy is called *potential* energy and is due to the gravity of the planet. For the stone to get this energy, we must lift it off the ground. In other words, we must do work. In this case we say that we do work on the stone. Once the stone begins to fall the potential energy is transformed into *kinetic*

---

[17] A parcel of air is meant to represent a sample of air that we can follow as it moves. We can picture it as a cube of any size in the atmosphere. As this cube moves it will experience many changes.

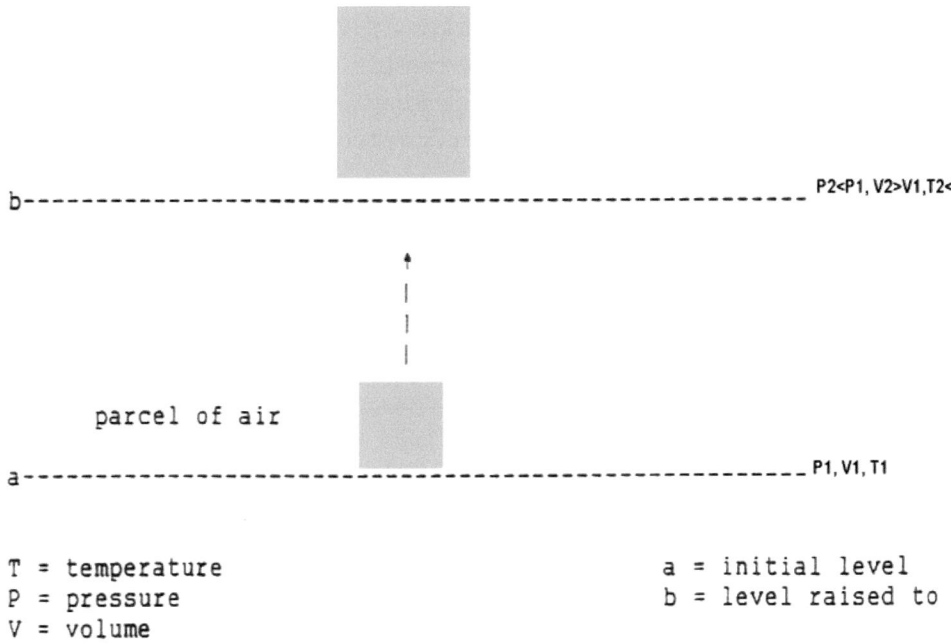

Figure 9: As the parcel of air rises it expands and eventually cools (see text for details).

energy. When it approaches the ground almost all its potential energy is transformed into kinetic energy, which is released to the ground upon impact. When a parcel of air is at the ground it has no potential energy to transform into kinetic energy. As such, it will not move unless some energy is given it. For example, when a car is set in motion it is the chemical energy of the fuel that is transformed into kinetic energy. When it comes to a parcel of air, motion can be achieved by absorption of radiant energy from the sun or through conduction with a warm area; the parcel warms and begins to rise. We can thus see that, while in general energy is the ability to do work, this ability comes in many forms.

Let's go now back to the rising parcel. Since the parcel is pushing the surroundings, it is doing work on the surroundings, thus it is transferring some of its energy to its environment. As a result, some of the parcel's kinetic energy is spent. Less kinetic energy means slower motion and, according to the definition of temperature, slower motion means lower temperature. *Therefore, as the parcel rises it cools* ($T_2 < T_1$). To reach this conclusion, we have to assume that as the parcel rises and expands, it does not mix with the environment. In a sense, we assume that its hypothetical boundaries can stretch, but they hold the air inside them and do not allow it to interact with the air outside, like the air inside a balloon. Of course, we know that in the atmosphere this cannot be true, but for some applications this is an acceptable assumption. Because of this assumption, this cooling is called *adiabatic cooling* from the word αδιάβατος. This Greek word refers to an area or place or object, which has not been walked upon. It signifies the fact that there has been no interaction with outsiders. If we reverse the arguments, and now consider a sinking parcel, we will conclude that as air sinks from a higher level, its temperature increases (i.e. higher temperature at lower levels than at higher levels). This is called adiabatic warming. Adiabatic cooling and warming are of paramount importance in meteorology. They are responsible for the cooling of the air's temperature as we climb the lower atmosphere. The air close to the surface gets warm, rises and sets off convection patterns of rising and sinking motions. It follows that because of convection the temperature of the air will decrease with height, which is opposite to the 'expected' distribution of temperature with height. However, there are limits to this. As we discussed above, when the parcels rise and cool they also slow down. Thus, at some point they will have no more kinetic energy, and they will be to cold (and heavy) to move any higher. At this point, the rising motions will cease. This boundary marks the end of the lower layer of the atmosphere (the troposphere), where all weather takes place and it is called the *tropopause*. Its average height is about 12 km (8 miles).

Above the tropopause there are no rising and sinking motions, thus the effect of convection is gone and temperature is now going to do what we expect from theory. In the stratosphere temperature begins to increase with height. Helping here is ozone, which is found in the stratosphere and as we discussed earlier absorbs most of the ultraviolet part of solar radiation. This increase in temperature continues up to about 50 km and at that level the air is warm and thin enough to be able rise again. This is similar to what is happening in the troposphere, and again it results in a steady decrease in temperature with height for up to about 90 km where it becomes too cold for rising motions. From that level on temperature increases again with height and actually it increases very rapidly. We are now in the thermosphere. This rapid increase, is as we mentioned earlier, compounded by the striping of electrons from the very energetic x-rays and gamma rays which creates ions and releases heat in the environment. As with the tropopause, the boundary separating the stratosphere from the mesosphere is called the stratopause and the boundary separating the mesosphere from the thermosphere is called the mesopause. The temperature distribution of temperature in the atmosphere is shown in Figure 10.

Analogies and contrasts

First and paramount, is that Aristotle's four principles and four elements (aether excluded) are at the heart of what we call today *Thermodynamics*. Formally, thermodynamics is defined as the study of equilibrium states of a system, which has been subjected to some energy transformation. More specifically, thermodynamics is concerned with transformations into mechanical work and of mechanical work into heat. This basis is extended to Atmospheric Thermodynamics. Our atmosphere is

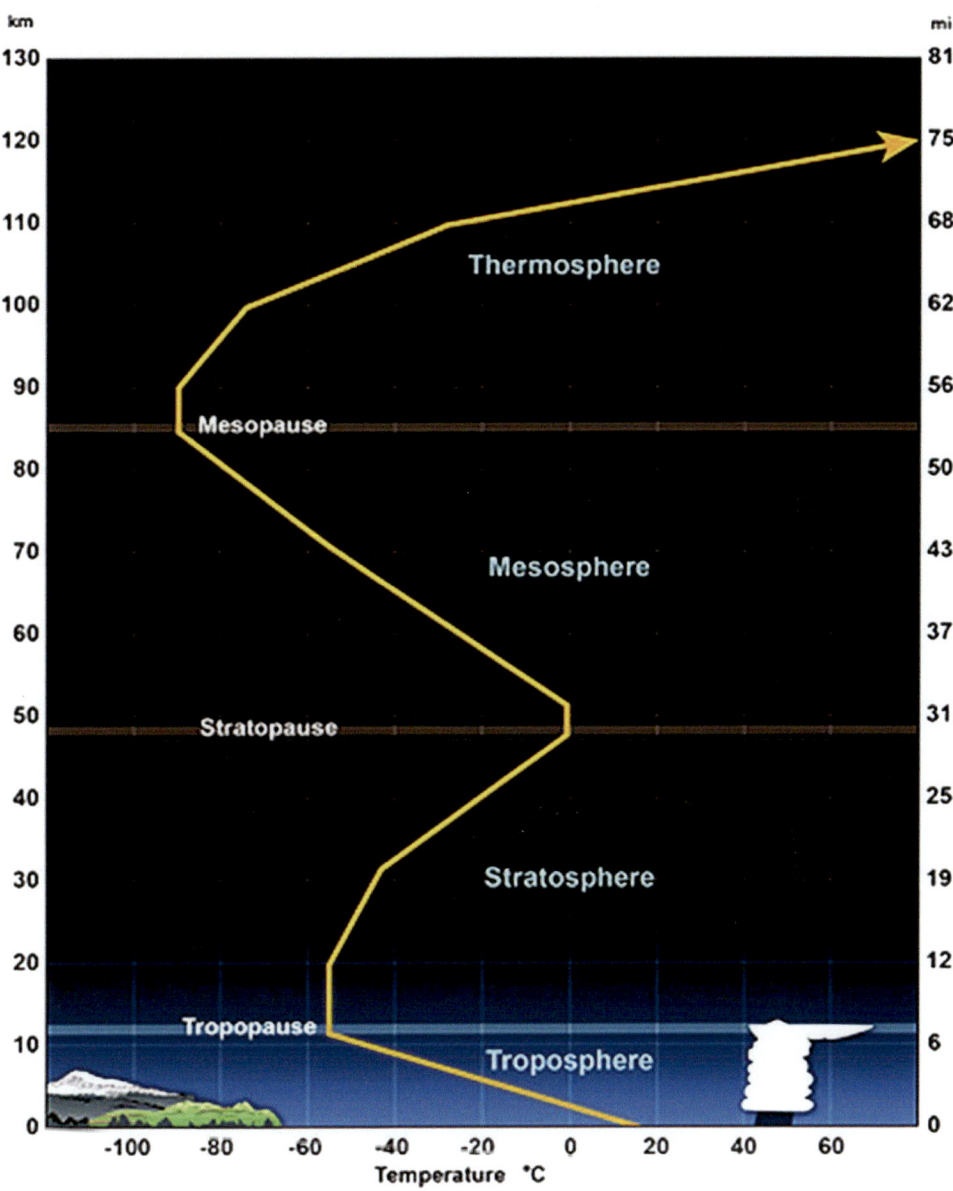

Figure 10: The vertical structure of the atmosphere. The yellow line shows how temperature changes with height (courtesy of NOAA).

Key points of meteorology now, part 1

1. 99% of the atmosphere is nitrogen (78%) and oxygen (21%). Water vapour is a variable gas. It is introduced by evaporation of liquid water when it is warmed up by solar radiation and it is removed via precipitation.
2. The atmosphere has weight. This weight over a given area defines the atmospheric pressure over that area. If the atmosphere were motionless, the pressure would be the same at all points at any level in the horizontal (i.e. the pressure field in the atmosphere is stratified).
3. When the Sun heats the surface, the air in contact with it warms up. Warm air is lighter than its surroundings and as a result it rises to higher levels.
4. The rising air is replaced at the surface by colder air rushing in from the surroundings. This closes a loop in the vertical which is called convection. This loop of rising warm air and sinking cold air, occurs at all scales from small (local) to planetary scales; lake breezes, sea breezes, Hadley cells.
5. Convection destroys the stratification of pressure creating high and low pressures in the horizontal. Once we have that, motion of air is initiated from high to low pressure areas. This is what we call wind. By definition, rising or sinking motions are not qualified as wind.
6. Moist air is heavier than dry air.
7. As the air rises (sinks) it cools (warms).
8. Water is found in three states: vapour, liquid, and ice. When we go from vapour to liquid and liquid to ice, heat is released into the environment. In the opposite way, heat is taken from the environment.

basically a two-component system. One component is dry air and the other is water existing in vapour and possibly one of the condensed sates (liquid or ice). Atmospheric Thermodynamics describes all processes involving transformation of water to vapour, vapour to water, water to ice, ice to water, vapour to ice, and ice to vapour, with all their heat exchanges, and all processes that result in formation of precipitation, and many other phenomena in the atmosphere, including weather forecasting. Atmospheric Thermodynamics is one of the most important parts in weather and climate modelling. Given the absence of measurements and technology at the time when Aristotle (and other Greek philosophers and scientists) were speculating on these principles, such insights are nothing less than extraordinary.

Now recall Aristotle's major inferences in chapters 1- 4, 9:

1. The universe is geocentric with all other 'heavenly bodies' revolving around Earth.

2. The distribution of the four elements and aether in Aristotle's universe shown in Figure 3.
3. The Sun warms the planet and water evaporates.
4. Notion of moist and dry exhalation. Atmosphere as a two-component system.
5. Dry exhalation rises because it is warmer. Moist exhalation is heavier and sinks below the dry exhalation.
6. Warm air is lighter, cold air is heavier.
7. Warm air rises, cold air sinks.
8. When it is cold, water vapour condenses to produce water and clouds.
9. Condensation releases heat.
10. Notion of convection.
11. The Hydrological cycle.

Inference 1, is clearly wrong.

Inference 2, has some truths to it. Most of the surface of the Earth is covered by water, moisture and clouds are found closer to the surface than higher in the atmosphere, at higher levels the air is dry, and at the top of the atmosphere 'fire' reigns (it's indeed very hot there).

Inference 3, is clearly correct.

The notion of exhalation, while ingenious, is not directly related to convection. Dry air is permanent in the atmosphere, and it is not exhaled by land. True enough, the air today was 'exhaled' from the planet due to volcanic activity, but it is not being exhaled anymore. Water vapour is a variable gas present only when water evaporates. Nevertheless, the notion of exhalation and his two-component atmosphere system, provided Aristotle with the means to elaborate on rising motion and then convection.

The remaining of inferences are mostly accurate. The Sun warms the surface and water evaporates. As we explained, warm air is lighter, warm air rises, and as it rises it cools. When it gets cold enough, water vapour condenses and releases heat (called latent heat of condensation), and eventually falls down as precipitation. The only statement that is wrong is the one made in Chapter 4: 'From these two kinds (of exhalation) only the second is able to rise because it is warmer whereas the first one being moist stays on the surface because of its weight.' We now know that moist air is actually lighter than dry air, But, at the time of Aristotle, there was no way for anyone to know the chemical composition and molecular weights of water and dry air. Moisture in the air, simply meant 'more stuff', therefore more weight.

The notion of convection is rather accurate for both smaller scales and hemispheric scales. It looks as if Hadley (recall Figure 8) had some help from the writings of Aristotle some 2,300 years earlier. As Aristotle says in chapter 9: 'Therefore, if earlier writers were attaching some secret meaning to 'Oceanus' (ocean), they probably meant this river which flows in a circle around the Earth'. In addition, according to Hadley's simple circulation (Figure 8), the air at the surface rushes from the North Pole in the northern hemisphere and from the South Pole in the southern hemisphere. This will make the surface winds in the northern hemisphere mostly north winds, and in the southern hemisphere mostly south winds. In Book B, as we will see next, Aristotle indeed states that most winds are north in the northern hemisphere or south in the southern hemisphere. He goes on to argue that this is due to the fact that the north and the south are the only regions the sun does not visit. His insight that the north and south receive less radiation is astonishing.

Before we conclude with Book A, it would be interesting to make some parallels between convection and Aristotle's exhalation. Convection is the paramount process in weather. As we have discussed several times, convection is a cyclic motion of warm rising air and cold sinking air. Exhalation is the major 'axiom' of Aristotle in Μετεωρολογικά. Basically, it is air coming from inside the Earth. While moist exhalation can be thought of as evaporation, dry exhalation does not exist. However, insofar as both convection and exhalation are associated with heating from the Sun, the sequence: rising motion → cooling → clouds → precipitation, led Aristotle to describe the hydrological cycle, vital for all life on Earth: a major achievement, as well as the atmospheric circulations of warm and cold air.

On now to precipitation...

Back to Aristotle's *Meteorologica*

As we have already mentioned, Chapters 9-12 deal with precipitation. For continuity, we repeat Chapter 9: 'Paramount to our discussion is the cyclic motion of the Sun.' Clearly Aristotle here refers to the daily cycle of the Sun. 'Since Earth is not moving, water on its surface evaporates, due mostly to the heat from the Sun, and it rises. When, however, this heat is gone, because it disperses at higher levels or because it is simply quenched by the air, its vapour in the rising air condenses and becomes liquid water and falls back to Earth. The exhalation from liquid water produces water vapour and its condensation makes a cloud. Fog is what may remain from the condensation into cloud. This is why fog is likely a sign of fine weather than of rain. Fog is something like a barren cloud. These changes in water phases follow the cyclic daily motion of the Sun. Truly, depending on how high or low the Sun is in the sky, the amount of vapour fluctuates.[18] We must think of such a phenomenon as a river moving up and down in a circle. When the Sun is closer (overhead), the stream of moist air flows upwards, and when the Sun is further (later in the day), it flows down. And this phenomenon is reproduced in the same order without exception. Therefore, if earlier writers were attaching some secret meaning to 'Oceanus' (ocean), they probably meant this river which flows in a circle around the Earth. To summarise: Moisture rises because it gets warm and falls back to Earth because it gets cold. Special names have been given to these processes and their varieties; falling small drops are called drizzle and bigger drops are called rain.'

In Chapter 10, Aristotle moves to the formation of dew and frost. 'From the amount of water vapour produced during the day, that which is not reaching the higher levels, because there is too much vapour, and heat is not able to rise all of it, remains close to the surface and cools at night. The liquid water forming then on the surface of Earth is called dew. If it is very cold, for example in winter, water vapour freezes before it can become liquid. We then get frost. Dew is formed when it is not very hot, so that the vapour does not dry out, and not very cold for vapour to freeze. Indeed, it is obvious that water vapour is warmer than liquid water (because it contains the heat that makes it rise), therefore it requires colder conditions in order to freeze. Both, dew and frost, occur when the sky is clear and there is no wind. The reason is that, if the sky were not clear, water vapour would not be able to rise, nor would be able to condense if it were windy. A proof that these phenomena are caused because water vapour does not rise enough, is that there is no dew on the mountains. One possible reason for that is that water vapour rises from evapourating water in hallow areas. In this case, the heat that is raising it gives the impression that it bears too heavy of a burden; it is not able to raise it and lets it fall back immediately. Another possibility is that wind, being more pronounced at higher levels, dissolves such a process.'

Aristotle concludes the chapter with a challenging paragraph that may be hard to follow. Here it is: 'Dew forms everywhere where south winds dominate and not when north winds are blowing, except in Pontus.[19] There the opposite takes place. The same reason that causes dew to from in warm places, but not in cold places, lurks here as well, but it involves an interaction of both cold and warm winds. South winds bring good weather. North winds bring bad weather because they are cold and quench the heat contained in exhalation. In Pontus, south winds do not bring enough warmth to cause significant evaporation. The north winds, however, being cold and heavy, squeeze whatever available heat from all sides. Such an action causes the heat to concentrate locally resulting in increased evaporation, which in turn allows for dew

---

[18] Thus, during the day the amount of vapour is a function of temperature.
[19] In the southern coast of the Black Sea.

to form when colder temperature take over. This happens in other places as well. The same principles apply to the observation that water from wells is warmer in the winter and colder in the summer.'

In Chapters 11 and 12 o Aristotle elaborates on the formation of the three distinct types of precipitation. In chapter 11 he says: 'From the region where clouds form, three types of precipitation emerge because of cooling: rain, snow, and hail. Two of these (rain and snow) are analogous to those that occur on the ground and are due to the same causes, the only difference being their quantity. As such, snow and frost are the same thing as are rain and dew, with the latter forming in less quantities than the former. Rain is due to the cooling of a great mass of water vapour, which accumulates over a large area and for a long time. By contrast, dew forms from a small amount of water vapour whose condensation takes a short time and usually covers a small area, as is evident for its little quantity. The same is valid for frost and snow. When the cloud freezes we get snow, when the water vapour freezes we get frost. That is why both frost and snow indicate cold season or cold regions, because water vapour would not be possible to freeze if it were not cold enough to overpower the presence of heat still present in it from the remains of heat which caused the evaporation of liquid water. Hail is formed in the upper region of the clouds. Hail is unique in the sense that no corresponding phenomenon exists in the vapourous region close to the surface. As we phased, snow in the upper levels corresponds to frost at the surface and rain to dew. But hail has nothing to parallel it. The reason will become clear next when we talk about hail in detail.'

In Chapter 12 he gets more involved: 'In our investigation on what happens during the formation of hail, we need to include those facts that appear indisputable, as well as, those that appear paradoxical. Hail is ice; however, hailstorms occur mostly in spring and summer, rarely during winter, unless it not as cold. In general, hail occurs in warmer climates, and snow in colder. In addition, it is illogical that water vapour will freeze in the upper atmosphere given that is not possible to freeze before first becoming liquid water, and liquid water is not able to remain suspended even for one a moment. The formation of hail is not analogous to the formation of large drops from tiny droplets. Those tiny droplets, because of their small size, are carried upwards inside the cloud and thus remain in the air (much alike small pieces of gold or soil floating on water). Given time, those tiny droplets join together to create large drops, which then fall to the ground, something that does not happen with hail because the crystals of ice cannot join as droplet can. It is obvious then, that the quantity of water corresponding to a hailstone, must have remained at those high levels for some time, otherwise the stones could not become as big.'

'Some suggest[20] the following cause for the origin of hail: Hail is formed when the cloud is forced to rise toward the upper atmosphere, which is colder because at that level the reflection of solar radiation from the Earth is not effective, and liquid water that may exist at this height freezes. That is why hail forms more often in the summer and in warm regions, because when the heat is great it thrusts clouds further up than normally. It so happens, however, there is less hail at higher elevations even though the opposite should be observed, as for example, in the case of snow which falls primary on higher elevations. We have seen many times clouds moving with great noise toward the surface,[21] terrifying those who hear them and see them as a sign of warning of some bigger catastrophe. Sometimes, when such clouds have been seen but without any noise, hail falls in great amounts, the stones are of unbelievable size, and their shape is not spherical, because their fall through the cloud did not take a long time, therefore, they were frozen near the surface and not, as some who we criticise here believe, far from the surface. Hailstones from high levels are more likely to break up during their long fall and to be shaped in smaller more spherical stones. Moreover, it is unavoidable that large hailstones be anything but the result of a strong influence of what causes freezing, because hail is ice, which is crystal clear. It follows then, that hail is not necessarily caused because the cloud was thrusted into the cold region of the upper air.'

'We will now talk about how warm and cold exchange roles and react upon each other. In warmer seasons, cold air, by the action of the surrounding heat, is concentrated and reacts causing the cloud to abruptly produce rain. The greater the contrast between warm and cold, the greater the 'squeeze' on the cold air, and the greater the speed of condensation, resulting in larger drops and in denser and more violent rainfall. This is exactly the opposite of what Anaxagoras says, who argues that indeed this phenomenon happens when the cloud reaches into cold air, whereas we believe that this happens when the (cold) cloud sinks into warm air, and the warmer the air, the more intense the rain. When cold air reacts even more to the pressure of surrounding heat, the water that is being produced freezes and produces hail. And this occurs when the freezing is faster than the fall of water. If water needs a certain time to fall, and is cold enough to freeze it in less than that time, nothing will prevent water from freezing in the air. The closer to the Earth, the more intense the freezing, the bigger the raindrops, and the bigger the hailstones because of the shortness of their fall. The large raindrops are not found in great numbers in this case. Hail is rarer in the summer than in spring and fall, but more frequent than in winter, because in the summer air is drier,[22] whereas in spring it is still moist and in autumn is becoming moist again.

---

[20] Aristotle refers here to Anaxagoras.
[21] Aristotle refers here to tornadoes.
[22] Keep in mind that Aristotle lived in Greece where summers are very dry.

What, in addition, contributes to the quick freeze is whether water is previously warmed, because then it cools faster. It is for this reason that people, when they want to freeze water faster, they display it in the sun. The inhabitants of Pontus, when they camp on ice to fish (by cutting a whole in the ice), they pour warm water on their fishing rods so that they freeze sooner; for they use the ice like lead[23] in order to settle down securely the rods.

The reason that in Arabia and Ethiopia rainfalls occur in the summer when torrential rains may fall frequently in the same day, rather than in winter, is the same. Again here, warm and cold exchange roles, and the great warmth of the country, causes colder air to be squeezed strongly and rapidly, which as we mentioned above, results in violent rainfalls.'

The above discussion concludes chapters 9-12. Arguably, the main inferences from chapters 10-12 (those from chapter 9 were discussed with those from chapter 1-4) are:

1. Clouds are produced by condensation of water vapour.
2. Water vapour amounts during the day depend on how high the Sun is (i.e. a function of temperature).
3. Dew and frost are formed because vapour at the surface condenses or freezes.
4. Both occur when the sky is clear and when the wind is calm.
5. There are three distinct forms of precipitation: rain, snow, and hail.
6. Rain and snow are analogous to dew and frost; their only difference is their quantity.
7. Large drops form from the union of tiny droplets which are carried upwards inside the cloud and they fall out as rain.
8. Hail is unique. No parallel phenomenon exists in the vapourous region close to the surface.
9. Hail usually occurs in warmer places and in spring and summer.
10. For hail to form, large drops have to remain at high levels for some time.
11. However, freezing of hailstones is not necessarily caused because the cloud was thrust into the cold upper air region. Hailstones may fall out of clouds moving close to the surface, and in this case, they are larger. The closer to the ground the more intense the freezing, and the bigger the hailstones.
12. Hailstones forming at higher levels are smaller at the surface because they break up during their long fall.
13. Warm water freezes faster than cold water.

---

[23] Ancient Greeks used molten lead to join the stones used on walls and pillars.

We will 'weigh' these inferences, but first another small tutorial.

Meteorology now, part 2

Clouds are the 'symptoms' of weather. When the weather is good certain types of clouds can be seen in the sky. When the weather is turning gloomy different types will be developing. Most weather calamities are associated, in one way or another, with clouds and precipitation. For this reason, we have to discuss the processes that underlie the development of clouds and precipitation.

We have already discussed that when a parcel of air rises it expands, does work on the environment, spends some of this kinetic energy, and cools. Cooling, however, is not the only change the rising parcel will experience. In order to describe what else is taking place inside the parcel, we have to introduce the definition of *humidity* and *relative humidity*. We have already discussed that water vapour is introduced into the air through evaporation of water from the surface of the planet. We have also established that when water vapour is introduced in dry air, it replaces some of the nitrogen and oxygen molecules, and that this modifies the weight of the air; contrary to the general feeling, it makes it lighter. The amount of water vapour in the air defines the humidity. Usually it is expressed as grams of water vapour in one kilogram of air. Unfortunately, humidity is often confused in weather reports on television, and even on our smart-phones (!), where humidity is given as a percent (for example, 70%, 90%, etc.). This practice is incorrect and we are not sure why it is practiced as such. The percent refers to relative humidity, which has a different meaning. Let us explain.

Consider a lecture room with seat capacity of N. At any given lecture the number of occupied seats may vary. In other words, the number of occupied seats may not always be equal to the capacity of the room. When all the seats are occupied the room has reached its capacity. If we think now of the occupied seats as water vapour molecules, and the capacity as maximum number of water vapour molecules a parcel of air can hold, we can clarify the difference between humidity and relative humidity. While humidity is the amount of water vapour in the air, relative humidity is the amount of water vapour in the air divided by the amount the air can hold. Thus, the relative humidity is a fraction whose numerator is humidity and the denominator is capacity.

Relative humidity = Humidity/capacity

In the lecture room example the capacity N is determined by the size of the room. In the atmosphere, it is the temperature that determines the capacity; the warmer the air the greater the capacity for water vapour. It follows, that the relative humidity could be high even if the numerator (amount of water vapour molecules in the air) is small, provided that the denominator is also small (which means cold temperatures). In other words, we don't get high relative humidity only in the summer when the temperature is high, but also during winter. However, in the summer high relative humilities feel 'sticky' because air at high temperatures can hold more water vapour.

The above definition of relative humidity suggests several possible ways of increasing the relative humidity. Two are special cases and the third is more complicated. The first special case, is to keep the temperature fixed (like in a room with a thermostat) and evaporate water in it (for example, with a humidifier). This provides water vapour in the air, thereby increasing its humidity. Since the temperature is steady, the capacity remains the same. Thus, the numerator increases and the denominator stays the same. The final result will be an increase in the value of their fraction (relative humidity). When the amount of water vapour reaches the maximum amount that the air can hold at that fixed temperature, the relative humidity becomes 100%. At this point the air is *saturated* with water vapour.

Now, what would happen if, after saturation is reached, more water vapour is added to the air? One might be tempted to say that the relative humidity will keep on increasing, but this is not what will happen. Once the air reaches its capacity, it cannot hold more water vapour. So, if we keep supplying the air with extra water vapour, this extra water vapour condenses and the resulted *liquid* water leaves the air, thereby returning the air back to saturated conditions. It is important to stress here that, just because there is condensation, it does not mean that there is no more water vapour in the parcel. Only an amount of water vapour equal to the increase of relative humidity above 100% is condensed. In a room, this condensation takes place on objects in the room, for example drapes, walls etc. That is why, if we over humidify, the room becomes damp.

The second special case for increasing the relative humidity is the inverse. The amount of water vapour is kept the same (numerator stays constant) but at lower temperature. This decreases the capacity (i.e. the denominator), and results in an increase in relative humidity. In winter, this will be equivalent, in the example with the room temperature, to opening the windows. Here again, after the air has cooled enough it will reach saturation. A further cooling will cause the relative humidity to rise above 100%, but as we discussed above this is not allowed. In this case, again, the extra vapour condenses to form liquid water or ice if it is too cold. According to this way, if we assume that during the day the amount of water vapour remains the same, then the relative humidity will be a function of the time of the day. Indeed, we usually register the lowest relative humidity at the time of maximum temperature (around 3 p.m.) and the highest relative humidity when the temperature is the lowest (just before sunrise). If the cooling during the night is strong enough, then the air might reach saturation, and with a further cooling the extra vapour will condense at the surface to produce dew. Thus it is not surprising that we usually observe dew forming during cool nights.

The third and more complicated way is a combination of both the special cases, when both the numerator and the denominator can change. For example, due to evaporation, an influx of vapour is supplied during the day, that could be moved around, thereby changing humidity beyond the two special cases. At the same time, temperature changes, which means the capacity changes. This will make the relative humidity fluctuate in a complex way. If we think that this process takes place at all space scales (local and global), and on a daily and annual cycle, we can imagine how complicated and unpredictable the processes of developing clouds and precipitation are.

Be this as it may, we can summarise that evaporation from water bodies supplies the air with moisture. This moist air will then either rise due to convection or it will be transported horizontally with the flow of the atmosphere. We know that rising motions cool the rising air. However, this is not the only way to cool the moist air. Moist air transported from the south to the north will also cool because it travels over colder areas. Thus, *both* rising and horizontal motion may cool the air.

Let us for now concentrate on air that rises. Consider a parcel of air, which at a given level (surface, for example) has a certain amount of mass and a certain amount of water vapour. This defines humidity. If we know the temperature of the parcel, then we know its capacity and therefore we know its relative humidity. Let's assume that to start with, the parcel is unsaturated (say, RH=70%). If the parcel begins to rise, then its temperature goes down. Because we assume that the parcel is adiabatic, it does not mix with the environment as it rises, therefore its mass and water vapour remain the same. Thus, its humidity does not change. However, because its temperature goes down, its capacity decreases and, therefore, its relative humidity increases. If the parcel's rising motion is strong, there will be a level where the temperature will be cold enough for the parcel to reach saturation. (Figure 11).

What is next? Just because the rising parcel reached saturation it does not mean that it is going to stop rising and cooling. Accordingly, the relative humidity of the parcel will keep on increasing. But, we have already discussed that when the relative humidity jumps above 100%, the extra vapour condenses into liquid water. Thus, just above the saturation level we will observe condensation and formation of tiny water droplets. A cloud now begins to form.

Here is a good place to introduce the *equilibrium vapour pressure*. The equilibrium vapour pressure refers to the amount of vapour over a water or an ice surface, when the water or ice is at equilibrium with the vapour over it. For example, assume that there is a pan full of water in a room. If the air in the room is not saturated with water vapour, then water evaporates and supplies the air with vapour molecules. Once the air reaches saturation, evaporation stops. The water left in the pan and the vapour in the air are now at equilibrium. The amount of water vapour in the air at this point has a certain weight given by the sum of all the water vapour molecules. This weight is pressing the surface of water in the pan. This defines a pressure, which we call equilibrium vapour pressure.

It follows, that the cloud bases occur just about where the parcel's relative humidity becomes 100%. We can now reason that, if the initial difference between the relative humidity at the surface were much lower than 100%, then the required cooling will also be great and as a result the cloud bases will occur at higher levels. Otherwise, the cloud bases will form at lower levels. Thus, the smaller the difference, the lower the cloud bases. In fact, if the difference is zero, then the cloud bases will form near the ground and we will get fog. It is also fair to say, that the lower the cloud base, the greater the possibility will be for the cloud to grow, simply because there is more room to grow (provided, of course that there is enough supply of water vapour). Consequently, the greater the development of the cloud, the greater the chance it will form precipitation. We are going to come back to this later.

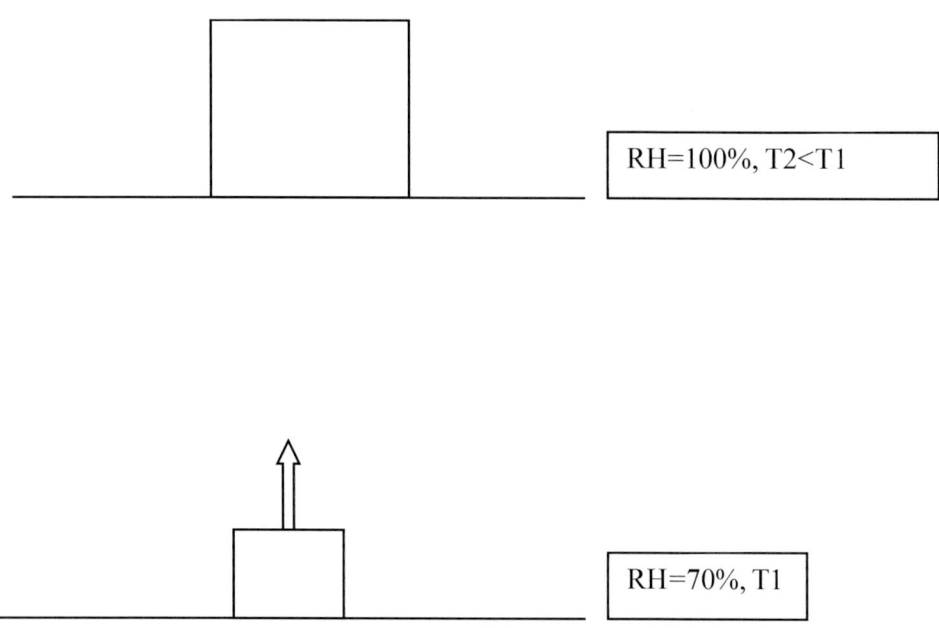

Figure 11: As the parcel rises it cools and its relative humidity increases.

We can conclude, that clouds form when the air is cooled below its saturation point. What is next? Let us try to imagine this rising motion in terms of movie frames (Figure 12). Frame #1 shows the cloud base forming. At this point, the extra water vapour has condensed into cloud droplets and the air's relative humidity has gone back to 100%. Here we will keep it simple and assume that the temperature remains above freezing as the parcel rises. Then, inside the parcel the two phases of water (vapour in small circles and liquid cloud droplets in large shaded circles) coexist, and (for the moment) are in equilibrium with each other. Here, this equilibrium is expressed as five water vapour molecules for every cloud droplet (in other words, the equilibrium vapour pressure is five pressure units). Equilibrium here means that, if no more changes occur in the parcel, water vapour and liquid water will coexist without increases or decreases in their amounts. But what would happen, if the parcel keeps on rising?

In frame #2 the parcel rises infinitesimally. Because of adiabatic cooling, i ts temperature decreases[24] and its relative humidity increases. Thus, its relative humidity jumps above 100% again. Because of the cooling, the parcel's capacity decreases and the parcel does not need the entire vapour it had in frame #1 in order to be saturated (or at equilibrium with the liquid phase). Let's say that this new equilibrium requires three water vapour molecules for one cloud droplet. Because of that, the extra vapour condenses and the cloud droplets grow. The same process occurs in all consequent frames. A continuous rising motion keeps the temperature of the parcel falling, which keeps its capacity for water vapour falling, which keeps its relative humidity jumping above 100%, which sustains more condensation, thereby developing the cloud and the droplets.

Will this procedure ever develop precipitation? Well, to make a long story short, as long as the droplets keep growing they become raindrops and eventually fall out of the cloud as rain.

The typical size of a cloud droplet is about 20 μm, and the typical size of a rain drop is 2,000 μm. This means, that a cloud droplet has to grow its volume one million times before it becomes a rain drop. At this rate, it is estimated than rain formation will take days. Observations, however, indicate that a cloud can develop and produce rain in about one hour. It follows, that something else must be present in rain formation.

A cloud can be either a warm cloud (having above freezing temperatures at all levels) or a cold cloud (extending to levels where temperature is below freezing). In warm clouds, only cloud droplets and water vapour exist. Our understanding of rain

---

[24] Note that because of the heat release due to condensation the parcel receives some heat. This does not mean that now the parcel will be warming as it rises but that the cooling rate will be lower than when it was not saturated.

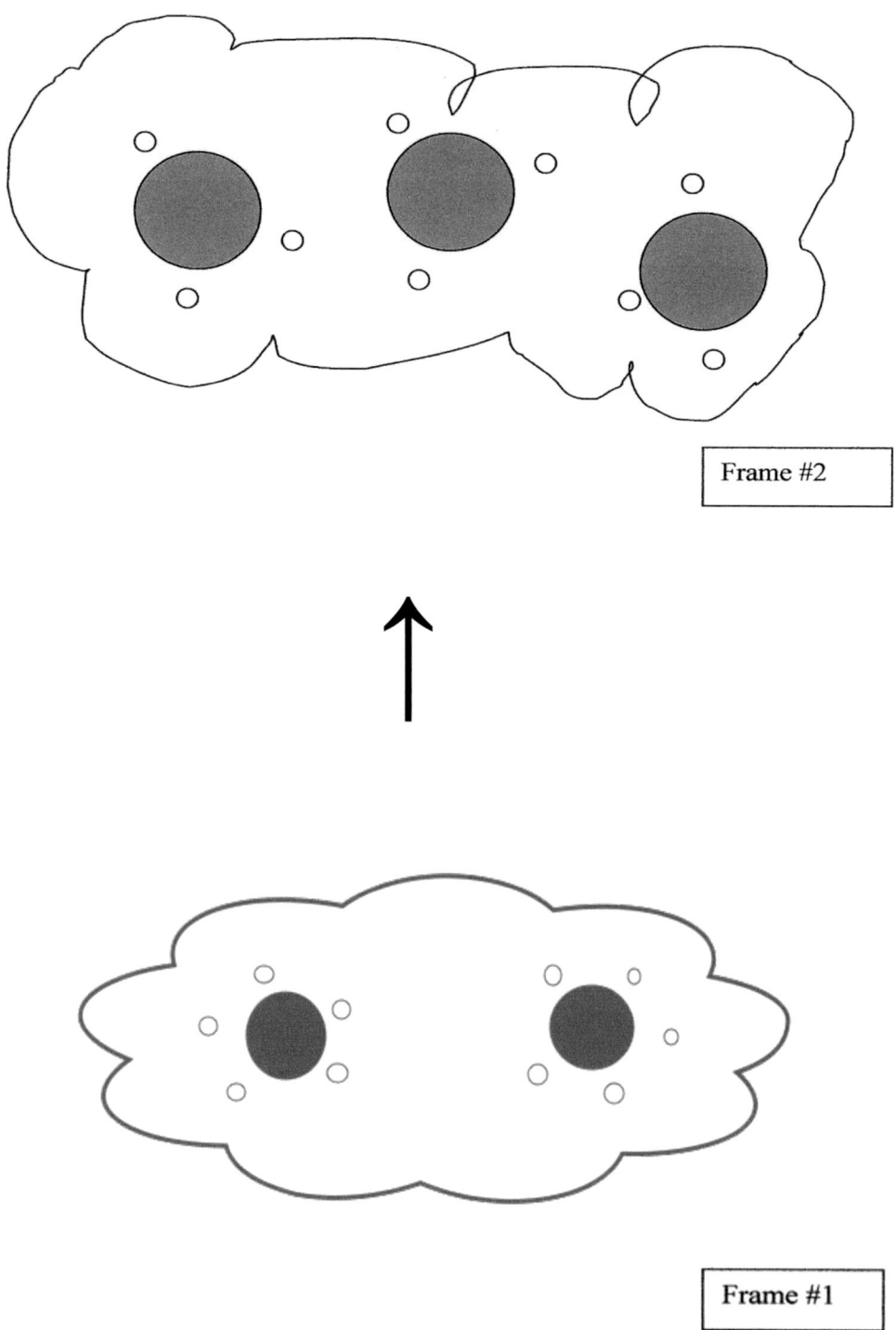

Figure 12: An illustration of how a cloud grows when the air that has reached saturation rises.

formation in warm clouds is that cloud droplets grow by a mechanism which involves collision between cloud droplets (known as coalescence). When cloud droplets form, they don't all have the same size. Local fluctuations in water vapour result in a non-uniform droplet size distribution. The larger droplets fall faster and overtake and collide with smaller droplets. In many of these collisions, the smaller droplets merge with the bigger droplets thereby causing the bigger droplets to become even bigger, which in turn, through the same mechanism, increase their chance to grow even further and become raindrops. When the drops are big enough they may overcome the updrafts (rising motion) inside the cloud and fall out as rain. When these updrafts are very strong, the drops remain in the cloud longer and as result they grow bigger. This explains why the raindrops are larger when the clouds are big.

In cold clouds the situation is very different and more interesting. In cold clouds, *all three phases coexist.*[25] Cloud droplets and ice crystals will now compete for vapour. Which one will win? Let's consider the situation frame by frame again shown in Figure 13. In frame #1 the cloud base has just formed. At this point, the extra water vapour has condensed into cloud droplets and ice crystals, and the air's relative humidity has gone back to 100%. Inside the parcel the three phases of water (vapour, liquid and solid ice) coexist and are in equilibrium with each other. If we look carefully into this frame we will observe something very interesting. Even though the three phases are at equilibrium with each other, the distribution of vapour is not uniform inside the developing cloud. The presence of ice crystals causes the distribution of water vapour to become uneven. There are more water vapour molecules around the droplets than around the ice crystals. The reason for this 'unfairness' is the equilibrium vapour pressure.

Physics dictates that the equilibrium vapour pressure over ice is less than that over liquid water. The number of water vapour molecules required in order for ice and water vapour in the air to be at equilibrium is less. This fact, will result in a picture inside the cloud where the water vapour molecules are arranged so that there are more vapour molecules around the droplets than around the ice crystals. This in turn will create 'mini' high pressures (H) around the droplets and 'mini' low pressures (L) over the ice crystals.

Recall now what happens when high- and low-pressure areas are formed. Always in this case there is motion from high to low pressure. The same is going to happen inside the cloud between the 'mini' high and low pressures. Accordingly, in frame #2 we will observe that water vapour molecules are swept from around the droplets (high

---

[25] In simple terms, this is because even though at high levels temperature may be below freezing, the atmospheric pressure is much lower than at the surface. This allows for all phases to coexist.

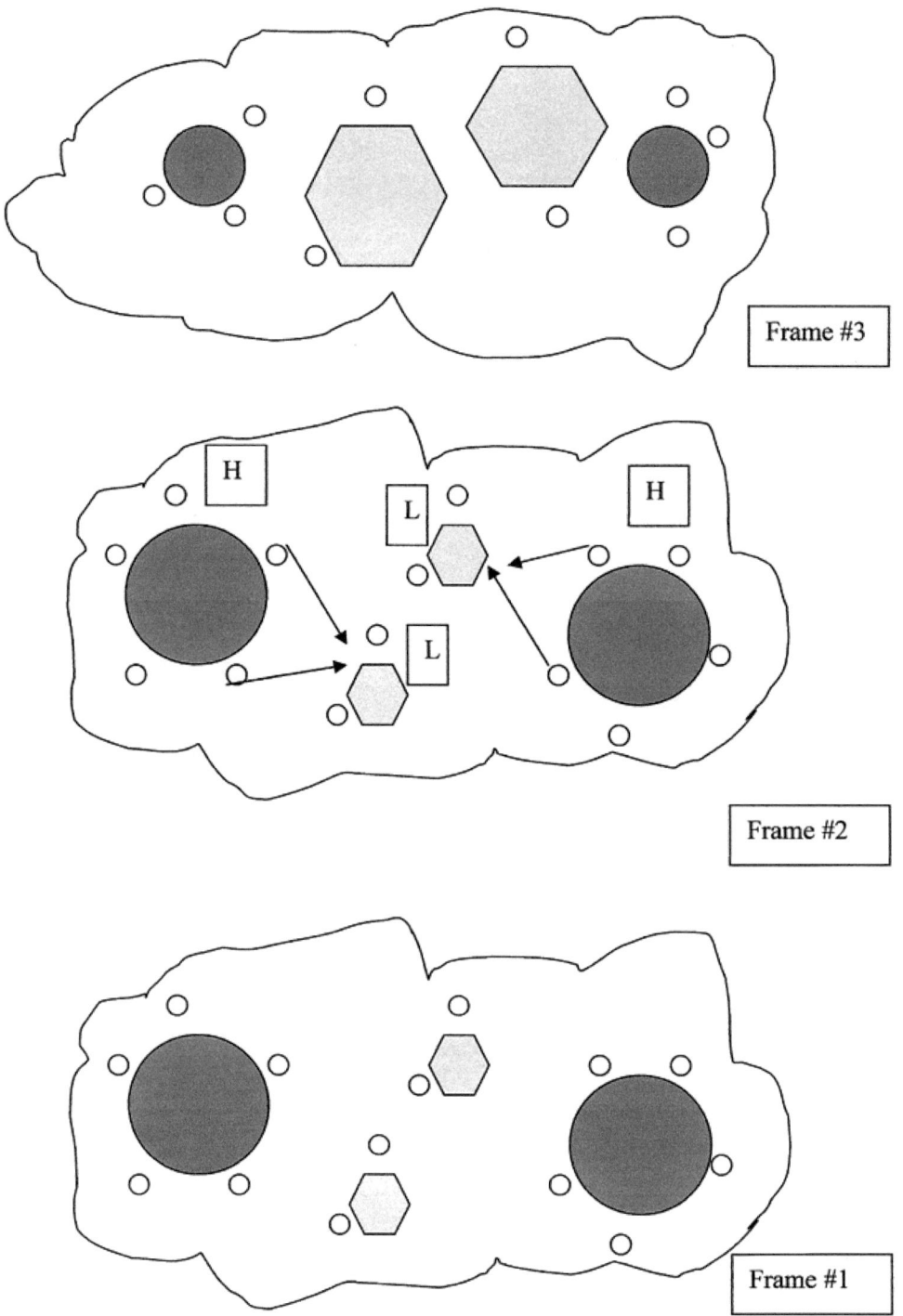

Figure 13: Illustration of the ice-crystal process.

pressures) towards the ice crystals (low pressures). Since the ice crystals are receiving water vapour molecules, they now have more than they need to be at equilibrium, whereas the droplets have less than they need. Thus, in frame #3 we will notice that the ice crystals use the extra vapour to grow, whereas the droplets evaporate to return to equilibrium with the air around them. This process continues and the ice crystals grow at the expense of the surrounding liquid droplets. Interestingly, even though there are much less ice crystals than droplets in the air, the crystals win. This process is called the ice-crystal process or the Bergeron process, in honour of the Swedish meteorologist Tor Bergeron who suggested this mechanism.

Ice is a crystal made of water vapour molecules arranged on a hexagonal lattice. Because of that the most basic ice-crystal shape is the hexagonal prism, which includes two hexagonal basal facets and six rectangular prism facets (Figure 14). As the crystal is blown through the cloud, water vapour molecules from around diffuse on it. Because the corners of the crystal stick out a bit more than the other points, water vapour molecules have a greater chance to diffuse to the corners. This causes the corners to grow faster, and to stick out even more, which causes more water vapour molecules to stick on them. This positive feedback is called *branching instability* and it causes corners to grow into branches and random bumps in the branches to grow into side-branches. The final result is a beautiful and complex structure: the snowflake.

The branching instability and random motion of the water vapour molecules cause each snowflake to be different from any other snowflake (Figure 15).

Other possibilities exist. For example, ice crystals may collide with the very cold droplets.[26] In contact, this cold water losses energy and freezes, thereby sticking to the ice crystal. This icy structure is called graupel or ice pellet. However, whatever forms in cold clouds, may by the time it reaches the ground become rain, if below the cloud the temperature is sufficiently high to melt the icy products. Thus, even

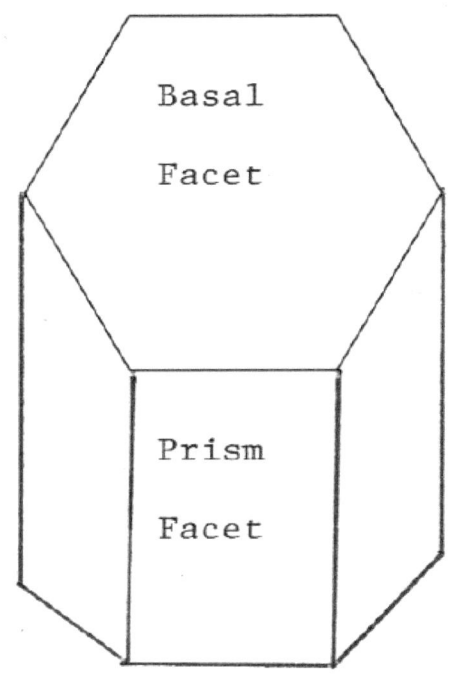

Figure 14: The basic geometry of an ice crystal.

---
[26] These drops and droplets are called supercooled. Remember that even at lower freezing temperatures liquid water can still exist because the pressure is very low at higher levels.

Figure 15: A snowflake.

though the ice-crystal process may initially produce snow, it often delivers rain on the ground. Of course, depending on the vertical structure of temperature at a given day, many types of precipitation may occur. For example (Figure 16), assume that snow forms in the upper levels and begins to fall through a melting layer. Depending on how deep this layer is, some or all of it may melt. If below this melting layer and close to the surface, is a freezing layer, some of the rain may freeze to give ice pellets, rain, and snow on the ground. Again, depending on the depth of this freezing layer, the rain may not freeze, but may cool enough to freeze on impact with the ground. In this case, freezing rain may be added to the menu. There are days (usually February or March) that in several locations in the United States such vertical temperature structures (a freezing layer close to the surface and a melting layer above it) occur. A mixed bag of products ranging from snow to ice pellets, to rain and freezing rain may precipitate. Those are the days to stay home and avoid driving! Note, however, that anything that is not rain or snow, is simply a by-product of them as modified by temperature. Ice pellets, freezing rain, sleet, and the like, are not new types of precipitation.

Let us now discuss a form of precipitation that is not rain or snow or a byproduct of them. This third and distinct form of precipitation is *hail*. A necessary ingredient in the formation of hail is a cloud with very strong updrafts (rising motion) in its lower part. Initially, as the cloud develops to levels where the temperature falls below 5°F (-15°C), ice particles begin to form. As these ice particles fall, they accrete supercooled liquid droplets and grow to become graupel. This graupel will serve as an *embryo* on

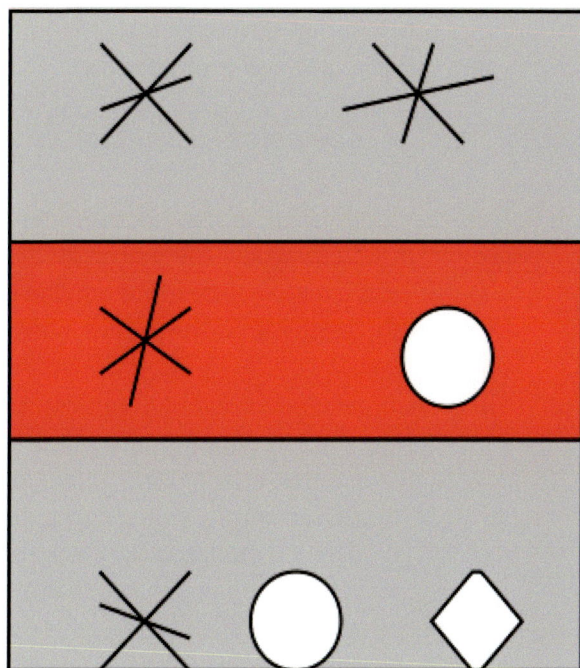

Figure 16: All or some of the snow formed in the upper cold layer may melt if it travels through a middle, melting layer. Then some raindrops may freeze going through the lower freezing layer thereby bringing a mixture of types of precipitation on the ground.

which a hailstone will grow. Under normal conditions, this graupel will either fall out of the cloud as ice pellets or snow. When, however, the updrafts at the *lower* part of the cloud are very strong, they force the falling graupel back to higher altitudes, where the process of growing by collecting supercooled water is repeated This way, the graupel particles grow rapidly to hail size. While in the updraft motion, the growing hailstone may reach a size that balances the opposition of the rising motion. At this point the hailstone 'floats' in the updraft. This provides the hailstone with extra time to grow into larger sizes. When the stone grows enough to overcome the updraft it falls out of the cloud.[27]

So far we concentrated mostly on precipitation forming when the air is rising and cooling. As we mentioned before, however, the air can also cool when it is transported in the horizontal from the warmer south to colder north. This alone, however, will

---

[27] The growth of hailstones is characterised by two regimes; the dry and the wet growth regimes In the dry regime, as the supercooled drops freeze at the surface of the hailstone, they release heat (in the same way as condensation releases heat). If during this action, the surface of the stone remains below 32°F (0°C), then the accreted supercooled liquid remains frozen and the growth is dry. On the other hand, if the stone's surface temperature rises above 32°F, the accreted liquid does not freeze. In this case, water diffuses into the porous regions of the stone. The hailstone still grows, but now the growth is a wet growth. Inside hailstone producing clouds, the turbulent motion and inhomogeneous distribution of liquid cause the hailstones to collect supercooled water at varying rates. Thus, it is possible that both regimes may take place during the formation of hail. In this case, the hail may acquire a layered structure (similar to that of an onion), each layer representing a change from the dry to wet growth regime.

not produce precipitation. Some vertical motion should be present and even then, we usually get light precipitation (rain or snow).

Key points of meteorology now, part 2

1. Moist air is a mixture of dry air and water vapour. The air cannot hold any amount of water vapour. The maximum amount of vapour the air can hold depends on the temperature of the air. The higher the temperature, the more the water vapour the air can hold. Thus, warmer air can hold more water vapour than colder air.
2. This maximum amount is called *capacity*. However, just because the air has some temperature, it does not mean that it holds the maximum amount. It may hold any amount from zero to maximum. The amount of vapour the air holds at a given time is called *humidity*. The ratio of humidity/capacity defines the relative humidity (RH). When the ratio is equal to 1, the air is said to be *saturated* with water vapour (or RH=100%).
3. When unsaturated air cools, its capacity decreases, therefore, the relative humidity increases. It follows that sustained cooling will cause the air to saturate.
4. A further cooling will result in relative humidity reaching values above 100%, which will correspond to an excess of water vapour above the limit of RH=100%. This is not allowed. As a result, the extra vapour condenses to produce liquid water when temperatures are above freezing or ice when they are below freezing.
5. Cooling may happen in three ways. When the air at the surface cools by loosing radiation (for example, at night), when the air rises, and when warm air is transported in the horizontal over colder regions. In the first case, dew or frost occurs. In the other two cases, clouds begin to form.
6. As the rising air keeps on rising the cloud grows and eventually may develop precipitation: rain by growth of the cloud droplet due to sustained condensation and collision/coalescence between cloud droplet, snow in a more complicated process involving growth of ice crystals at the expense of cloud droplets, or hail by an even more compilated process requiring very strong rising motion and floating of growing large ice aggregates.

Analogies and contrasts

Recall Aristotle's main inferences from chapters 10-12:

1. Clouds are produced by condensation of water vapour.
2. Water vapour amounts during the day depend on how high the Sun is (i.e. a function of temperature).
3. Dew and frost are formed because vapour at the surface condenses or freezes.
4. Both occur when the sky is clear and when the wind is calm.
5. There are three distinct forms of precipitation: rain, snow, and hail.
6. Rain and snow are analogous to dew and frost; their only difference is their quantity.
7. Large drops form from the union of tiny droplets which are carried upwards inside the cloud and they fall out as rain.
8. Hail is unique. No parallel phenomenon exists in the vapourous region close to the surface.
9. Hail usually occurs in warmer places and in spring and summer.
10. For hail to form, large drops have to remain at high levels for some time.
11. However, freezing of hailstones is not necessarily caused because the cloud was thrusted into the cold upper air region. Hailstones may fall out of clouds moving close to the surface, and in this case, they are larger. The closer to the ground the more intense the freezing, and the bigger the hailstones.
12. Hailstones forming at higher levels are smaller at the surface because they break up during their long fall.
13. Warm water freezes faster than cold water.

Inference 1 is clear and accurate. It is mentioned directly in several places (for example in chapter 9). It is also mentioned indirectly earlier in chapter 3.

Inference 2 is more or less accurate. The higher the Sun, the higher the temperature, the stronger the evaporation, the more the amount of vapour in the air (i.e. the higher the humidity). Note, however, that here Aristotle does not make a distinction between humidity and relative humidity (like the smart phones!).

Inference 3 is accurate.

Inference 4 is also accurate. The surface cools at night because heat is emitted from the surface into the atmosphere. If the sky is clear, this heat escapes more rapidly than if it is cloudy. Thus, the surface and the air in contact with the surface will get cooler faster in clear nights. This increases the possibility of getting dew or frost. For dew or frost to form at the surface, the air at the surface should be the coldest. When the surface is cooling it is reasonable to assume that the air close to the surface will be the coldest because the air above receives some of the escaping heat and absorbs it. If, however, there is a strong wind, mixing occurs and the coldest air may not be at the surface.

Inference 5 is also clear and accurate.

Inference 6 is only marginally correct. It is correct only in the fact that vapour condenses to form cloud droplets in a warm cloud. Aristotle missed the ice-crystal formation of rain. Aristotle never mentioned ice and water coexisting inside a cloud.

Inference 7 is a true statement, as it describes a major component in rain formation in warm clouds: Aristotle mentions that droplets in the cloud join to create large drops. This reflects the collision and coalescence process to produce large drops that fall out as rain.

Inference 8: so far so good!

Inference 9: true. This is when we expect the heating on the ground to be enough to result in strong rising motion, which builds the cloud well into the higher levels of the atmosphere.

Inference 10: partly correct. Recall our discussion above 'While in the updraft motion, the growing hailstone may reach a size that balances the opposition of the rising motion. At this point the hailstone 'floats' in the updraft', and compare it to Aristotle's *'It is obvious then, that the quantity of water corresponding to a hailstone, must have remained at those high levels for some time, otherwise the stones could not become as big'*.

Inference 11: Here Aristotle loses his argument to Anaxagoras, who correctly suggested (chapter 12) that *'Hail is formed when the cloud is forced to rise toward the upper atmosphere, which is colder because at that level the reflection of solar radiation from the Earth is not effective, and liquid water that may exist at this height freezes. That is why hail forms more often in the summer and in warm regions, because when the heat is great it thrusts clouds further up than normally.'* In his effort to explain the observation that there is less hail in higher elevations, (which may not necessarily by true), and to 'fit' into his argument the fact that hail is falling out of *'...clouds that that move with great noise toward the surface...'* (obviously referring to tornadoes), he resorts to the concept that *'large hailstones are the result of a strong influence of what causes freezing, and it is not necessary for the cloud to be thrusted into the cold region of the upper air.'*

Inference 12: Not true

Inference 13: this phenomenon, is today known as the *Mpemba effect* named after a Tanzanian student who discovered, in cookery classes in the early 1960s, that a hot ice cream mix freezes faster than a cold mix. This effect is questioned, as it is not clear whether 'freezing' refers to the point at which water forms a visible surface layer of ice, or the point at which the entire volume of water becomes a solid block of ice, or when the water reaches 0°C (its freezing temperature). One popular explanation is that evaporation of the warmer water reduces the mass of the water to be frozen. Evapouration is endothermic, meaning that the water mass is cooled by vapour carrying away the heat, but this alone probably does not account for the entirety of the effect.

We need to discuss here a notion that permeates several of Aristotle's explanations. Those explanations involve *antiperistasis*. The term comes from the Greek ἀντιπερίστασις, formed of ἀντί ('against') and περίστασις ('standing around), and means 'resistance to anything that surrounds or besets another'. Antiperistasis, in philosophy, is a general term for various processes, real or contrived, in which one quality heightens the force of another, opposing quality. Otherwise stated, antiperistasis, is 'the supposed increase in the intensity of a quality as a result of being surrounded by its contrary quality.' We see Aristotle, invoking antiperistasis in several places. He directly or indirectly, assumes it as an explanation of 'paradoxical' events. Examples are: 1) the statement in the end Chapter 10 that in Pontus water of a well is warmer in the winter than in the summer; 2) the 'paradoxical' phenomenon that hail forms in the summer rather than in winter (Chapter 12); 3) his argument *'that large hailstones are the result of a strong influence of what causes freezing, and it is not necessary for the cloud to be thrusted into the cold region of the upper air'*; 4) his statement, again in Chapter 12, *'We will now talk about how warm and cold exchange roles and react upon each other. In warmer seasons, cold air, by the action of the surrounding heat, is concentrated and reacts causing the cloud to abruptly produce rain. The greater the contrast between warm and cold, the greater the 'squeeze' on the cold air...'*; 5) The reason that in Arabia and Ethiopia rains occur in the summer rather in winter. Again here, *'...warm and cold exchange roles, and the great warmth of the country, causes colder air to be squeezed strongly and rapidly, which as we mentioned above, results in violent rainfalls'*. We know very well that something lighter (warm air) cannot push something heavier (cold air); While Aristotle knows that, he is invoking antiperistasis in cases where he is not aware of a physical mechanism behind a phenomenon.

Today, many of these references have been explained through physics or through sensory adaptation. Take for example, the case of well water. Everybody knows that winter temperatures are much colder than summer temperatures. What everyone does not know is that, where usually well water is found, temperature stays pretty much the same all year, because the heat from the Sun cannot reach deep and therefore has no real effects on water's temperature. Thus, well water will be warmer than air temperature in winter and cooler than air temperature in summer. We humans, adjust to winter or summer air temperature and perceive it as 'normal'. We don't notice that well water temperature is nearly constant throughout the year. What we notice is the difference between air temperature and water. For example, assume that the temperature of underground water is 10°C (50°F). If the temperature outside is 30°C (86°F) you will feel that water is cold. In winter, if the temperature on the surface of Earth is 0 or below 0°C (32°F), well water with over 10°C (50°F) temperature and will definitely be felt as warm. Today it is well understood that the reason why well water appears warmer in winter than in summer is just another case of sensory adaptation. It may be that the observations in Pontus are simply microclimate.

Back to Aristotle's *Meteorologica*

The next two and final chapters in Book A, are not very relevant to weather. Chapter 13 begins with a mention about winds, but wind is discussed extensively in Book B and we will defer it for later. The remaining of chapter 13 is dedicated to rivers and seas and we will not discuss it here. Chapter 14 touches peripherally on what today we call climate change. Very few know that Aristotle acted not only as a philosopher but at the same time he went far away from supernatural explanations and beyond superstition. In fact, he was among the first scientists to try to present and hint on climate change both from a global and a regional perspective. His insights on this subject, we believe are stunning.

'It is not always that the same regions on Earth are dry or humid, but they change according to the appearance or disappearance of rivers. This is the reason land and sea are interchangeable. Land and sea are not fixed in space, but we find sea where it used to be land and where now there is sea, it will become land again. We need to admit that these changes follow some order and periodicity. The basic principle and cause are that Earth's interior experiences a time of maturity, like the bodies of plants and animals, and a time of decay; the difference being that the changes in plants and animals do not happen in specific part of them but they necessarily mature and decay as a whole. On the contrary, on Earth the changes take place in certain places under the influence of cold and heat, which depend on the course of the Sun. For the regions becoming dryer, it is the destiny of water springs to dry out, and as such big rivers become smaller and smaller and then dry out completely. Such changes affect the sea as well. In those regions, where fueled by the full rivers, the sea flooded the land, now it recedes leaving behind dry land. But the time will come when those places will be flooded again.'

'The physical change of Earth happens gradually and over appreciably long-time intervals compared to our length of life. As such, these phenomena pass unnoticed and whole nations are lost before they can preserve the memories of these changes from their beginning to their end. The most utter and sudden catastrophes are due to wars, pestilence, and famine. Some of these famines are of great scale, others of smaller scale. In the latter case, the disappearance of a nation may not be noticed, because some leave the country and some stay behind, until the land is unable to provide food to the few remaining inhabitants. From the time of the initial to the last departure, a long time has passed and nobody anymore remembers. The lapse of time destroys all record even for those last inhabitants that may still be alive. We also need to admit that in the same way, the date when first a nation settled in a land that was changing from wet to dry, disappears from memory. Because the change in the land occurred imperceptibly, and nobody remembers who were the first inhabitants, when they came, and what was then the state of the land.'

'This is what happened in Egypt. It is well known that this land becomes more and more dry and that the whole country is a deposit of the Nile. As the marshes gradually dried, neighbouring peoples settled there but with the passing of time they forgot their origins. It is certain that all the mouths of the Nile, except that at Canopus,[28] are human made, and that in older time Egypt was called Thebes. The higher regions where inhabited earlier than the lower ones. The places closer to the silt deposit by the river, were necessarily marshy for a longer time. Subsequently, those places change and become in turn more prosperous, because as they gradually dry acquire better quality. Those places, however, where initially there was a balance between dryness and humidity, became progressively worse.'

'This also happened in the land of Argos and Mycenae in Greece. At the time of the Trojan wars the land of Argos was wet and marshy and could only support a small population, whereas the land of Mycenae was thriving (which may explain its greater fame). Today the opposite is true. The land of Mycenae has dried out, but the valleys in the land of Argos are now arable. Given that the above changes can actually take place over regions of small sizes, we must admit that the same can happen over larger areas and even over whole countries.'

'The shortsighted ones believe that this kind of events lurk in some universal change, in the sense of a coming to be of the universe as a whole. This is why they maintain that, if the volume of the sea is reduced because of dry conditions, it is because today dry conditions are occurring in more places than in the past. There is some truth to this argument, because indeed today there are more places that used to be under water. However, if they examine this issue more carefully they will observe the opposite; they will find places where the sea has invaded the land. We should not, however,

---

[28] The farthest to the west, it is the only natural mouth of river Nile.

think that the reason for this is a universal change. For it is absurd that for small and trifling changes on Earth, to evoke the whole universe; after all the size of Earth is negligible compared to the universe. Rather, we should attribute the reason for these changes to the fact that they occur at regular time intervals. For example, while every year there is a winter season, after some determined long time interval a great winter will come accompanied by torrential rains and flooding.[29] This great flooding does not occur always at the same place. For example, the flood in the time of Deucalion[30] affected mostly the Greek area and especially Ancient Greece (Hellas) in the vicinity of Dodoni and of Achelous river, which often changes its course. In this area lived the Selli and those formerly called Graeki and now Hellenes. When, therefore, heavy rains occurred, we must assume that they lasted for a long time, and whatever is happening today with the rivers (some of which flow all the time and some don't), was happening then as well. Researchers support the idea that this is because of the presence of huge underground cavities, but we attribute it to the size of the mountainous regions, to their density, and the cold that dominates them (because these regions catch, store, and produce larger portion of water, whereas mountainous regions of small size or porous or stony regions see the water flowing away). We must believe, therefore, that the same is happening during great floods: in regions where water is accumulated, humidity increases thereby making them almost inexhaustible. But, as time goes on, the regions that dry out become more common, the wet regions less common, until we reach once again the beginning of the same cycle.'

'However, because some change in the universe must necessarily be happening, (without this being associated with coming to existence or with perishing; after all the universe is unchangeable), it should not be, in our opinion, that the same places are always flooded or dry out. This is proven by facts. Let us consider the Egyptians again, who are considered the most ancient of people. As we mentioned previously, their whole land is obviously the making of the river Nile and this is clear to anybody who takes a look around this country. An irrefutable proof is the Red Sea. One of their kings tried to make a canal there, in order to join the Nile with the Red Sea (for it would be a great profit if all this region became navigable; it is said that Sesostris was the first of the ancient kings, who attempted it, but he found out that the Red Sea was at a level higher than the land.[31] For this reason, Sesostris first and then Darius,[32] stopped the building of the canal. They were afraid that by mixing sea and river water, the river will disappear. It would therefore appear that this part of the world was once

---

[29] Remember again, Aristotle lived in Greece where snow is hardly seen, except high in the mountains.
[30] According to the myth, Deucalion was a king in the age of copper. When Dias (Zeus) decided to eliminate him, because he was a corrupted person, Deucalion, after his father's Prometheus advise, built a boat where he loaded the necessary, and thus, he and his wife Pyrra escaped the cataclysm.
[31] This perception was maintained until the 19th century. During Napoleon's expedition to Egypt (1798), the level difference was estimated to tens of meters!
[32] This is Darius A who reigned over Persia and Egypt from 521 to 486 BC.

an unbroken sea. For the same reason, in Libya, the region of Ammon[33] is lower and hollower than the land towards the sea. Clearly, silt deposits created a barrier which, as a result, formed lakes and dry land, and that with the passage of time, water in the lakes evaporated and now is gone. The same happened in lake Maeotis, where deposits by rivers were so significant, that the ships now have to be smaller to sail into it. As we have argued, this lake as well, is the making of the rivers and that at some time in the future it will all dry completely. Yet another example is Bosporus,[34] where due to river deposits we always observe strong currents. These strong currents come about because every time the current from the Asiatic shore throws up a sandbank, a small lake is formed, which then dries out, then a second sandbank, and a second lake, which then dries out, and so on. After many such repetitions, is obvious that the Bosporus strait must become some kind of a river, which in the end will also dry out.'

'It is, therefore, obvious that, since time is endless and the universe is eternal, neither Tanais[35] nor the Nile have always been flowing and that the region where they flow, was some time ago dry; for their energy has limits, while time does not. This is equally true for all rivers. But if indeed, rivers form and then dry out, and if the same regions on Earth are not always covered by water, it necessarily follows that the sea is also subject to similar changes. And if the sea is receding from some places but it advances over other places, it is clear that over the whole area of Earth, not always the same areas are sea or land, but the picture changes with time.'

The word 'climate' originates from the Greek verb 'kleino' (in Greek κλείνω). In the ancient thinking the word meant 'to be inclined', referring to the inclination of the sun's rays as we move from the equator to the pole, due to the spherical geometry of the Earth, which as we all know determines the climate to a large extent. Aristotle considered the Earth as being a sphere at the centre of the Universe and defined five climate zones based on the inclination of the sun's rays, concepts that have been put forward much earlier by Pythagoras and his student Parmenides (6th century BC). According to the knowledge of his time there were five climatic zones in a spherical world to temperate, two polar (frigid), two temperate and one torrid separating them. The torrid zone, called 'diakekavmeni zoni' was considered as being so hot that it was practically impossible to explore. The term 'tropical' came also from the Greek verb 'trepome' (in Greek τρέπομαι) which means the change in the course or the 'path' of the sun in his apparent movement from one hemisphere to the other hemisphere.

---

[33] In Ammon was the famous temple of Zeus, which Alexander the Great visited.
[34] Bosporus is a natural strait located in northwestern Turkey. It forms part of the continental boundary between Europe and Asia. It is the world's narrowest strait and it is used for international navigation. Bosporus connects the Black Sea with the Sea of Marmara, and, via the Dardanelles to the Aegean and Mediterranean seas.
[35] Tanais was an ancient Greek city in the Don river delta.

The change to 'tropi' (in Greek τροπή), in later centuries, gave birth to the term 'tropical' which refers to the tropical latitudes. The nomenclature of classical Greek understanding of different climates has been widely used by Renaissance mappers such as Homem's (1554) and the Blaeu (1648) atlases.[36] The persistence in time of the Greek (Aristotle's) idea of latitudinal climatic zones provided an interesting challenge to the explorers in the Renaissance who have already noticed that zonality was not the best representation of the various climates and this has been considered in Koeppen's classification of climates as late as in 1900 (Sanderson 1999).[37]

Clearly, Aristotle is talking about natural climate change occurring at all space/time scales due to intrinsic variability of the system. However, in statements like '*We need to admit that these changes follow some order and periodicity*' and '*On the contrary, on Earth the changes occur in certain places under the influence of cold and heat, which depend on the course of the Sun*', Aristotle may be making connections to what we know today as changes in the orbital characteristics of the Earth. Indeed, the regular cycles hypothesised by Aristotle have been specified 2000 years later by the Serbian geophysicist Milankovitch who was able to explain the recurrence of ice ages and the intermediate periods of warmth (interglacial) on the planet. His explanation referred to the so called 'geological cycles' which occur at approximate periods of 100,000, 40,000 and 20,000 years. Milankovitch has explained these quasi cycles to the long-term changes in the ellipticity of the earth's orbit around the Sun (about 100,000 years) as well as the cyclical changes of 40,000 years in the tilt of the Earth's axis and 20,000 year-cycle from the precession of the Earth's axis. The last interglacial period is the one we live now and it is characterised by higher temperatures and by the melting of ice sheets both in the high northern latitudes and in the Antarctica. This period of warmth will continue for more than 10,000 years from now but it coincided with the gradually accentuated warming caused by the excess emissions of greenhouse gases and other human activities. The human intervention to the environment has taken place always on Earth (for example the invention of agriculture during the Neolithic era), but it was so small before the so-called 'industrial revolution' that was passing almost unnoticeable and was completely absent on a global scale. After the industrial revolution a gradual destabilisation of climate was observed globally. The man's contribution relative to natural variability to these changes started from a few percent at the early 20th century and has accelerated exceeding 30% in the past decades. More on this can be found in Appendix II.

---

[36] Blaeu, W. J., 1648: *Nova Totius Terrarum Orbis Geographica ac Hydrographica Tabula*, auct: Guiljelmo Blaeuw. Excuedebat Gulielmus Blaeuw Amsterodomi. *Le Theatre du monde*, Amsterdami 1643-1646, Vol. 1, Clements. Library, University of Michigan, Ann Arbor, MI. Homem, L., 1554: (Mappamonde) *Lopo Homem cosmographo cauasero*, Cortesao, Armando: Portugaliae Monumenta Cartographica, Lisboa, 1960. Clements Library, University of Michigan, Ann Arbor, MI.

[37] Sanderson, M., 1999: The classification of climates from Pythagoras to Koeppen. *Bulletin of the American Meteorological Society* 80: 669–673.

BOOK B FROM ΜΕΤΕΩΡΟΛΟΓΙΚΑ

On winds

The first three chapters of Book B discuss water in association with the seas, such as salinity of water, and other issues. Chapters 7 and 8 discuss earthquakes. There is very little connection to weather phenomena in the 5 chapters and we will not discuss them here, except for distillation of water which has connections to hydrological cycle discussed in chapter 3. We will, however, go into chapters 4, 5, 6, and 9, where winds and other weather phenomena such as thunder, lightning, and storms are discussed.

In as early as 5th century BC Hippocrates stated that liquids including sea water could be made sweet by boiling. This view was further advanced by Aristotle in Book B, chapter 3 where he states: 'For the present let us just say that a certain amount of the existing sea water is always been drown up (because it gets warmer by the Sun) and is becoming sweet (not salty); and that it subsequently falls down as rain but in a different form from that which was drown up.' Aristotle went on to suggest an experiment where a vessel made of wax with its mouth fastened, in such a way as to prevent any water getting in, is put into the sea. What percolates through the wax is drinkable water, with the wax acting as a filter. That is why drinkable water is lighter than salty water because salt has been filtered out. That explains why ships that were loaded on rivers nearly sank in rivers when they were quite fit to navigate in the sea'.[1]

Aristotle starts chapter 4 as follows: 'We will now talk about winds. We have already discussed that there are two types of exhalation, one moist and one dry. The first we call vapour. For the other we don't have a general name, thus, in order to refer to it we must use a term that applies to one of its forms, for example some sort of smoke. The moist cannot exist without the dry, nor the dry without the moist. Whenever we speak of either we mean the one that dominates at the moment. As the Sun moving on its

---

[1] This predates the notion of buoyancy discovered by Archimedes more than 100 years later.

circular course approaches, it draws up with its heat the vapour and when it recedes the vapour gets cold and condenses into water, which falls back to the Earth.'

This last statement was also introduced in chapter 9 of Book A, and as we stressed earlier, it describes the hydrological cycle. According to Merriam-Webster dictionary, and in complete accordance with the Aristotle's four principles, 'weather' is the state of the atmosphere with respect to heat or cold, wetness or dryness, calm or storm, clearness or cloudiness. One can easily see that Aristotle's hydrological cycle relates to the Sun's cyclical motion, and the formation of clouds and precipitation. What remains from the above definition is 'calm or storm' (wind).

The ancient Greek philosophers and mathematicians were fixed into the 'romantic' idea that all phenomena or all maths can be explained starting with a few 'axioms'. An axiom is a 'truth' that needs no proof. For example, 'A straight line may be drawn between any two points' or 'if a=b and b=c, then a=c', are axioms. With his two exhalations, and his structured and inductive logic, Aristotle attempts to explain all meteorological phenomena. His moist and dry exhalations are his axioms in *Meteorologica*. And since the moist exhalation is connected to the hydrological cycle, according to Aristotle the dry exhalation must be behind the formation of wind. Here is what he states:

> 'Nevertheless, inside the planet there is great quantity of fire and heat, and the Sun does not just draw up the moisture that exists on the surface, but also it warms and dries the planet. Since we have two kinds of exhalation, one like vapour and the other like smoke, both must be necessarily generated. If the moist one dominates then rain is produced, and if the dry dominates it becomes the source and substance of all winds. This is true because the two types of exhalation are distinct and must differ. And, the Sun and Earth's heat not only can, but they must produce them'.

Thus, for Aristotle the two exhalations produce the air around us and then rain and wind are produced. Rain when the air is dominated by the moist exhalation, and wind when it is dominated by the dry exhalation. Thus, while rain and wind are two different entities, they co-exist. Here Aristotle also makes a connection to the rivers by arguing that wind must have a source. In his words, '...we don't consider every water that flows a river, even if there is a great amount of it, but only when the flow comes from a spring. The same applies to wind. A great quantity of air might be moved by the fall of a large object without flowing from a source'.[2]

---

[2] In this case it is not called wind.

In this chapter, based on his endless interaction between the two exhalations, Aristotle makes the following additional statements, based on the four-element theory and his own observations:

1. More rain occurs during night than day, and in winter than summer.
2. Great spatial and temporal variability in rain and droughts is not surprising.
3. During rain wind ceases and after rain wind rises.
4. Dry exhalation is the cause of wind.
5. Most winds come from the north or from the south, and they have a somewhat oblique direction.
6. Winds originate in the higher levels.

Let us consider each statement in contrast with today's knowledge of these issues. Aristotle makes the first statement based on the reasoning that to produce rain, the moisture must be condensed (thus must be cooled). Since the temperatures are colder at night than during the day, and in winter, chances are it will rain more during nights and in winter. In general, this is not a correct statement. We have discussed that precipitation may develop when the air at the surface rises and cools due to adiabatic cooling. If the heating of the air at the surface (which causes the air to rise) is strong, then the cloud grows in the vertical and precipitation develops. And when is the heating strong? Obviously during the day and in the summer. This way of producing precipitation, is not the only way. Precipitation is also associated with the so-called low-pressure systems that develop from the interaction of cold air coming from the north and warn air coming from the south. When the two air masses collide, there is an action whereby the cold air, being heavier, goes under the warm air and lifts it from the surface. This is what happens when the bathroom door is opened after a warm shower was taken. One's feet get cold as colder (and heavier) air from the room next to the bathroom enters the bathroom and lifts the warmer air to the ceiling. This action creates a three-dimensional wave in the atmosphere. In order to picture how this wave looks and what it will do, imagine the following experiment.

Fill up your bathtub with water. Then place your hand at the bottom on one end and vertically raise your palm rapidly. This action will cause water to splash upwards exactly where the lifting of your hand is taking place. Now, what do you think you will observe on the other end of the bathtub? The motion of your hand creates a wave in the bathtub. Since this wave cannot move the bathtub, water at the other end slides over the bathtub and spills on the floor. Thus, in this example, we see that the generated wave is characterised by a lifting of water on one end and sliding over the bathtub on the other end. Now think of water in the bathtub as the warm air coming from the south and your hand as the cold air coming from the north. Where they meet, the cold air being heavier goes under the warm air and lifts it up. This creates a wave in which the warm air tries to push the cold air aside. This is not possible because warm air is lighter and as a result, having nowhere to go, the warm air slides over the colder air to the north. This action continues as the cold air keeps on going under the warm air (this is like repeating the lifting of your hand many times as if you wanted to remove all the water from the bathtub). This process results in what we call the *warm front* (where the warm air from the south is sliding over the cold air from the north), and the *cold front* (where warm air is lifted vertically). Along both the fronts the warm air is *cooling*. Along the warm front, because it travels over colder regions, and along the cold front because of adiabatic cooling. Thus, it is possible that precipitation will develop along both fronts. These low-pressure systems develop everywhere on the planet and then they travel around the globe. Thus, they can deliver precipitation day or night.

The second statement, which is based on his discussion in Chapter 4 of Book B, is in principle correct, and it resonated with Aristotle's comments on 'climate change' discussed at the end of Book A. He states: 'Since the exhalations occur constantly, sometimes in stronger and in greater quantities than other times, we always get clouds and winds. Because sometimes it is the moist exhalation that dominates and sometimes it is the dry, some years are wet and some are dry and windy. It also happens, that sometimes, excessive rains or droughts are frequent and affect whole countries, or some regions more than others. It is also often possible that, while one region is experiencing drought conditions, regions around it get the normal amounts of rain expected in a given season. Contrariwise, it is possible that a place in the middle of a country, which normally experiences little rain or even drought, receives large amounts of rain. The explanation for this is the following: while in most cases it is natural to observe the same phenomenon over most of the area of a country, for the reason that nearby places have a similar position with respect to the Sun (unless there is something special about them),[3] dry conditions will sometimes replace naturally occurring moist conditions in one area, and vice-versa, as they may move around driven by the wind.' Even though Aristotle makes use of his 'axioms' here, the great variability of precipitation and drought in time and in space is captured clearly.

The third statement cannot be generalised. Often wind and rain are very much associated.

The fourth statement is not correct because Aristotle was not aware of atmospheric pressure and its spatiotemporal variability. More on this will follow later.

---

[3] Notion of microclimate.

The fifth statement touches on a very important point. Aristotle's explanation that the winds come mostly from the north or the south[4] is based on the following reasoning: the cyclical motion of the Sun on a daily basis and on an annual basis, is from east to west. Therefore, the north and south receive less radiation, meaning they are colder, meaning that the clouds will form on the sides of the east-west trajectory. Thus, precipitation will fall to the north and to the south. After precipitation falls, the heat from the Earth and the Sun dries out the surface and dry exhalation ensues. Since winds must have a source and since winds are caused by dry exhalation, they must then come from the north or the south. Aristotle also states that the actual north or south direction is somewhat oblique or slanted because the air is subject to the rotation of the universe, and thus is influenced by it.

While it may be true that most surface winds are from the north or the south, it is not because of dry exhalation to the north or to the south. If we recall Figure 8 and its discussion, one should expect that, because of the Hadley circulation cells, winds in the northern hemisphere will blow at the surface from the north pole to the equator, and those in the southern hemisphere will blow from the south pole to the equator; hence north or south winds! But, there is a 'catch' here: the planet rotates. This is something that was not realised in the Aristotelian view of the universe. The rotation of the planet causes a force called the Coriolis force, the effect of which is to deflect a motion to the right in the northern hemisphere and to the left in the southern hemisphere. Given this, air that is moving from the north pole toward the equator, will be deflected to the right of its course, thereby making the direction of the wind oblique (more like northeast). Similarly, the direction of the wind in the southern hemisphere will be move towards the southeast. Thus, Aristotle was correct in mentioning that the direction of the wind is oblique, but this was based on observation and not on correct physical reasoning. The connection to some kind of rotation is, however, ingenious.

The simple one cell Hadley circulation model is unstable. There are many reasons for it being unstable. First, the planet's surface is not uniform. The distribution of land and water is not homogeneous. Second, it is subject to the earth's rotation, which will twist it differently at lower and higher latitudes. The net result is that each single Hadley cell will break into three smaller circulations. Be that as it may, this break-up will result into three general wind belts near the surface of the planet: a northeasterly flow between the equator and $30°N$ latitude, called the trade winds, a southwesterly flow between $30°N - 60°N$, called the prevailing westerlies, and another northeasterly flow between $60°N$ and the north pole, called the polar easterlies. There are thus

---

[4] Note that Aristotle lived in the area of the Aegean Sea, where winds are bounded by topography and blow from north or south directions. More on the discussion on this, later in this chapter.

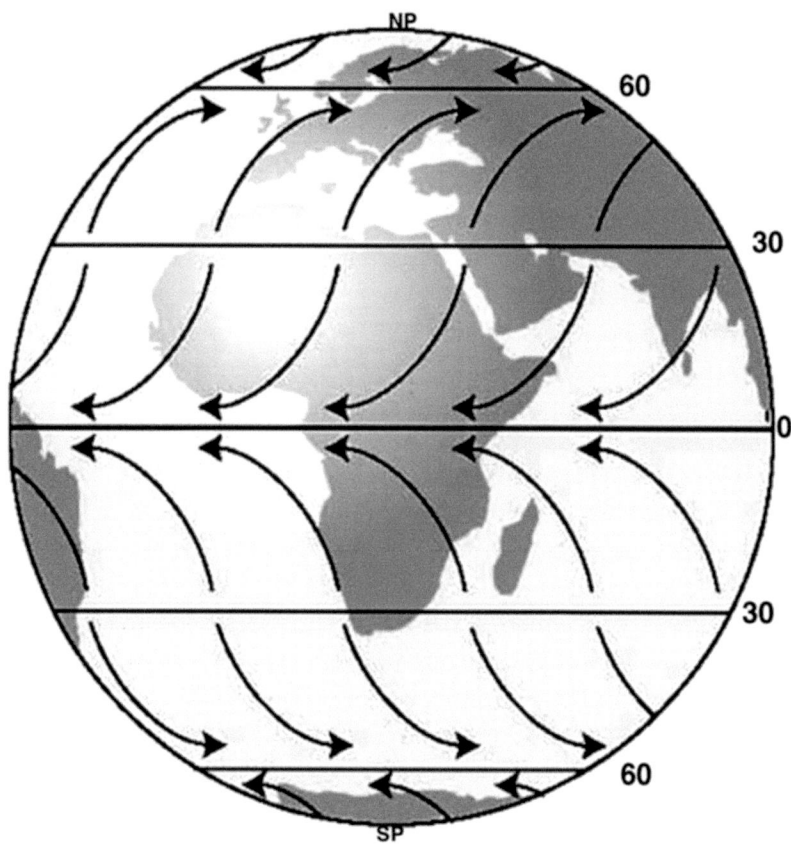

Figure 17: The three general wind belts at the surface of the planet.

plenty of north and south components in the wind direction. A similar pattern is established in the southern hemisphere (Figure 17).

Aristotle's final discussion in this chapter (sixth statement) refers to whether wind originates from below or above. The discussion is not very clear, but the point Aristotle is trying to make here is that wind must originate from above. He argues that, while the *material* cause of the wind is the exhalation from the bottom, the *driving* cause is from above. He points to the fact that before we feel the wind at the surface, its presence above is betrayed by the movements of clouds. Thus, wind must originate at higher levels. Recalling Figures 7 and 8, and the associated discussion, it is true that horizontal motion (wind) originates at higher levels after air is moved from lower to higher levels due to convection (note that rising or sinking motion is not wind). However, the reasoning Aristotle provides is not concise.

In summary, statements 1, 3, 4, 6 are not justified, statement 2 is correct, and statement 5 has some truth to it, but for the wrong reason. Aristotle's dry exhalation as the cause of wind does not lead him to a correct description of the genesis of winds. It is interesting to note here that Aristotle never talked about atmospheric pressure. We believe that the reason, is that in general, putting regard in as truth for a long time, becomes something like an axiom. The early thinkers, did not 'feel a weight on their head'. Clearly then, for them the air was not excreting any force on their heads. Such axioms prevailed as long as there was no way to measure 'things' or technological advances to make new observations. Had Aristotle come up with the concept of atmospheric pressure, and if he had any evidence that the Earth is not the centre of the Universe, but it spins on itself while it revolves around the Sun, the state of science would have been advanced much earlier. Nevertheless, Aristotle, displayed an unparalleled way of organising thoughts and observations and through logical arguments arriving to inferences that stood (and many still stand) the test of time.

In chapter 5, Aristotle continues on the topic of winds. At the beginning of the chapter, Aristotle notes that winds are weak or do not blow when it is too hot or too cold, and that stronger winds occur sometime between winter and summer. This statement is partially correct. On the average, winds around the globe are weaker in the summer and stronger in the winter. Here is the reason: The tropics receive more or less the same amount of radiation throughout the year. That is why in the tropical regions the temperature remains more or less the same. The poles on the other hand receive minimal sunlight in the winter and more in the summer. Thus, there is a stronger temperature gradient (difference) between the poles and the equator during the winter months. The temperature gradient affects the pressure gradient. Since the wind strength depends on the magnitude of the pressure difference, it follows that on the average there will be stronger winds during winter. Locally, however, in both space and time, because of temperature gradients between land and water, or because of strong storms, tornadoes, hurricanes, or special topographical characteristics, winds can become strong at any time of the year.

Aristotle then describes the most dominant wind in the area of Greece, the *etesians* in more details. The name etesians derives from the ancient word ἐτησίαι, which means 'yearly'. They are very strong, dry north winds blowing in the Aegean Sea, from July to mid-September, being the strongest in August.

The Aegean basin extends some 600 km in the north-south direction, with a narrow straight at the latitude of Athens that is less than 200 km wide. The red colour on Figure 18 shows elevation above 700 metres. The figure clearly shows a 'channel' created by the mountains between Greece and Anatolia. The mean surface pressure field for July-August in the period 1981-2010 is shown by the black lines. This field is dominated by a high-pressure system centred in Azores (pressure values over 1020 millibars) and extending to the Balkan area and a low-pressure system centred over the valley of central Iraq and extending to Turkey (pressure values less than 1002 millibars). The air with the high-pressure system circulates clockwise, and the air with the low-pressure system circulates counterclockwise. This combination, together with the effect of friction, results in a northerly wind over the Aegean (shown by the black arrow) which is remarkable for its persistence,[5] and the strength of which is further intensified by the 'channel'.

On average, the etesians are stronger during the day and weaker during the night. However, there are times when etesians blow continually for days. An interesting point is that this arrangement is actually connected to a much larger scale pattern involving the summer monsoons of the Indian subcontinent. An unusual characteristic of the etesians in the Aegean, and one which causes some confusion in terminology, are the periods of enhanced etesians. Within the narrow straight of the Aegean, periods of gale winds (≥ 6 in Beaufort scale) are relatively frequent. But when reference is made to the etesians, it is often not clear whether the author intends the term to mean the general monsoon background or whether it is to be used in the more restricted sense to the periods of unusually strong north winds.

The etesians were known very well in ancient times for their periodicity and strength and the ancient Greeks, as well as other ancient civilisations in Eastern Mediterranean, depended on seafaring. Aristotle describes these winds, their timing (after the summer solstice) and their strength, with great accuracy. Again, with accuracy, and based on years of observation and experience, Aristotle describes the south winds which replace the etesians after the winter solstice. The south winds are not as fierce as the etesians, in fact they often go unnoticed. As we have mentioned, Aristotle did not involve atmospheric pressure in his descriptions of weather phenomena. As a result, he based the explanation of the genesis and strength of the etesians and of the south winds on the interaction of the Sun and as well as on the two exhalations during summer and winter, which, as we have mentioned already, leaves a lot to be desired.

Aristotle next speculates on some very interesting thoughts. According to the prevailing wisdom of the times,[6] 'A region is inhabitable unless shadows appear toward the north all days in the year. Our north region (the only one Aristotle was interested in), is habitable only up to a distance close to the Tropic of Cancer, because close to the tropic shadows disappear. Exactly on the line of the tropic, the Sun is at its zenith and there are no shadows. When we cross it toward the south, the shadows

---

[5] This constancy if often compared to the constancy of the trade winds.
[6] O.J. Tricot, *Les Meteorologiques, nouvelle traduction et notes*. Paris, Vrin, 1955.

ON WINDS

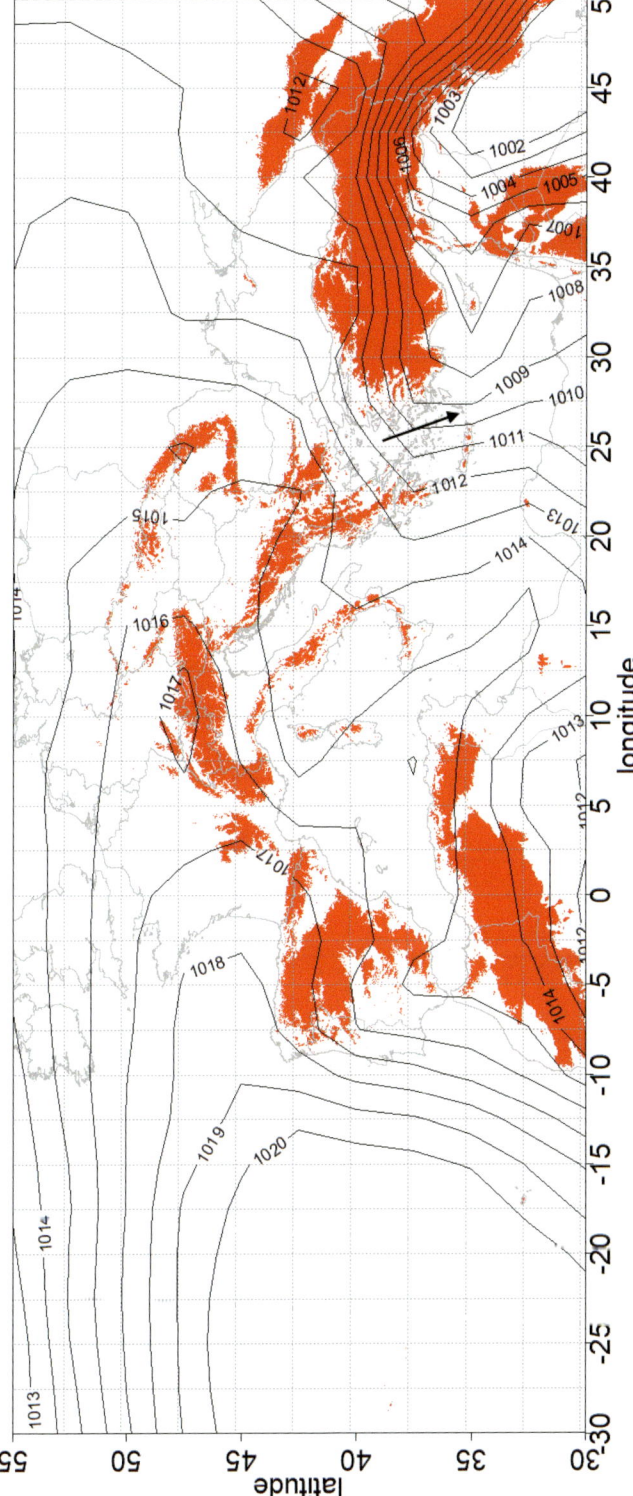

Figure 18: The setup behind the etesians. The red colour shows elevation above 700 meters, which clearly forms a 'channel' by the mountains between Greece and Anatolia. The black lines show the mean surface pressure field for July-August, and the black arrow resulted direction of the wind (see text for details).

are directed towards the south'. In other words, south of the Tropic of Cancer is too hot for people to live. Aristotle also believed that regions are inhabitable north of a latitude (most likely referring to the Arctic Circle), because it is too cold. Aristotle thus considers regions with mild, moderate climate habitable. He then goes on to argue that the habitable zones may have limitations with regards to latitude, but they may extend around the planet. As Figure 19 illustrates, the habitable regions would then be, those within ABCD section in the northern hemisphere, and those within EFGH section in the southern hemisphere. 'Indeed', Aristotle argues, 'the line connecting the Pillars of Heracles[7] to India, is much longer, in proportion 3 to 5 to that extending from Aethiopia to Maeotis and the northmost Scythians,[8] as far as such matters are considered accurate, given our experience with travels by sea and by land'. Just for comparison, Figure 20 shows population density on Earth today. Indeed, most of the population lives between the tropic of cancer and the arctic circle in the northern hemisphere.

This is not valid for the southern hemisphere which mostly water, a fact not known by the ancients.

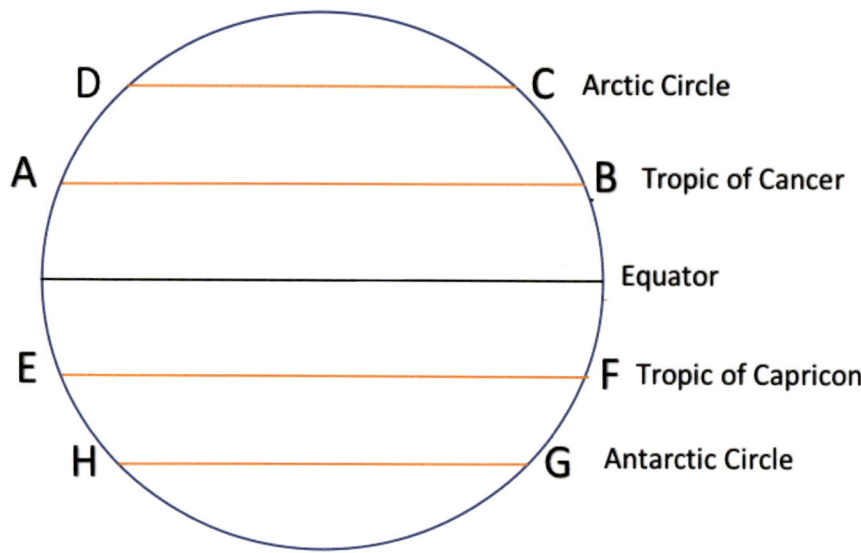

Figure 19: A simple diagram depicting the habitable zones (ABCD and EFGH) according to Aristotle.

At the end of chapter 5, Aristotle goes on to conjecture the existence of Antarctica. 'Now since there must be a *region* bearing the same relation to the south pole as the

---
[7] Straits of Gibraltar.
[8] Central Eurasia.

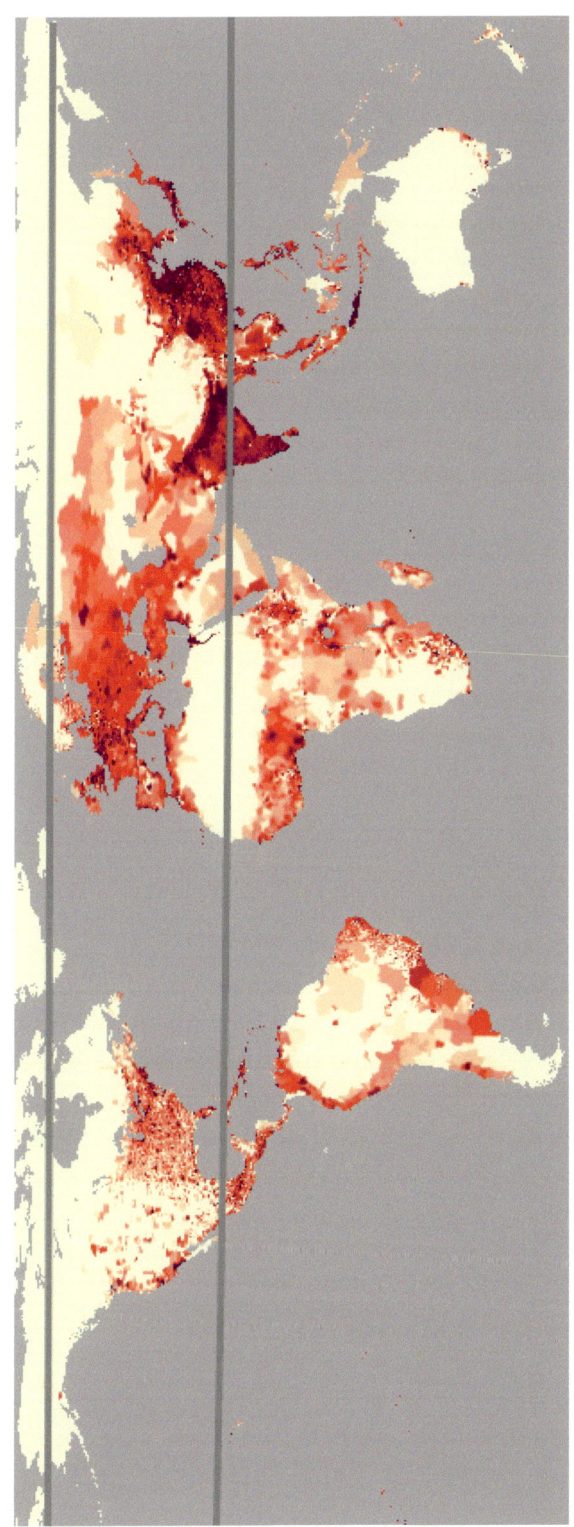

Figure 20: Population density on Earth. The grey horizontal lines are the Tropic of Cancer and the arctic circle, respectively. The darker the red colour the higher the density. Source: coolgeography.co.uk

place we live in bears to our pole, it will clearly correspond in the ordering of the winds as well as in other things. Thus, just as we have north winds, there must be corresponding winds coming from *their pole*.[9] But this wind cannot reach us since even our own north wind does not reach the limits of the region we live in. It is also clear that our south wind is not the south wind blowing from the south pole…but from the Tropic of Cancer…' The *region* he suggests was the first conjecture of the existence of Antarctica. In honour of this conjecture, we now have the Aristotle Mountains in Antarctica.[10]

Interestingly, his thoughts are very much aligned with the average picture of the general atmospheric circulation shown in Figure 17.

In chapter 6, Aristotle describes the main local winds (other than the etesians) prevailing in the area. He gives their names, the direction from they were blowing, pointing out that the characteristics of each wind depend on where they are coming from. This chapter is a matter-of-fact listing of the winds in the area of ancient Greece, and we will not discuss further. Instead, we will move to the last chapter of Book B (chapter 9) and to chapter 1 of Book C where Aristotle talks about stormy weather, lightning, and thunder.

Stormy weather

Chapter 9 of Book B begins with Aristotle's explanation of thunder and lightning. Here is his take on the subject: the two exhalations combine to produce clouds. During condensation, some of the dry exhalation is trapped inside the cloud. This trapped exhalation may be squeezed out as clouds collide. The collision is accompanied by a noise, which is thunder. This phenomenon is compared to the phenomenon referred to as the 'laugh or threat of Hephaestus';[11] and the sound produced in burning wood, when the exhalation, after it is compressed, is rushed to the flame.[12] The ejected wind is usually inflamed and burns with a thin and faint fire. This is lightning produced at the point where the exhalation is coloured in the act of its ejection. It comes after

---

[9] Here Aristotle uses the word Άρκτος, which is usually translated as 'pole'. However, Άρκτος in Greek means 'bear'. Aristotle is referring to Ursa Major, a constellation in the northern sky, in the northern hemisphere. Arctic literally means 'near the bear, northern'. Aristotle by referring to 'their pole' or in the ancient text 'της εκει αρκτου', is conjecturing an 'Antarctic' in the south hemisphere.
[10] *Aristoteles* is a crater on the *Moon* bearing the classical form of *Aristotle's name*.
[11] Greek god of fire.
[12] Indeed, inside the wood, there are tiny pockets of fluids, such as water and sap. As the wood burns, the fire heats these fluids and causes the fluids to first boil and then vapourise into steam. This steam gets trapped in the pockets and begins to exert pressure on the surrounding wood. Eventually, the wood gives way. The snap, crackle, or pop sound you hear is the wood splitting along a crevice and releasing steam into the fire. This is why wet wood snaps, pops, and crackles much more than usual. More steam is produced by the heat.

the collision and after the thunder, but we see it earlier because sight is quicker than hearing.

Aristotle then mentions that Empedocles and Anaxagoras have a different view: Empedocles claims that fire exists in the cloud in the form of intercepted Sun's rays, and Anaxagoras maintains that this fire is aether, which he calls 'fire' and which has descended from above. In those two hypotheses, the gleam of the fire is lightning, and the hissing noise of its extinction is thunder.[13] That, Aristotle, states, will indicate that lightning comes before thunder, which does not make sense. With Anaxagoras he argues that, if this fire is aether drawn down from the above, then we should see lightning not only when it is cloudy but also when the sky is clear. With Empedocles, he insists that the explanation is careless, because if it were true, then thunder and lightning must have a separate and deterministic causes assigned to them, meaning that they pre-exist and they just emerge.[14] In this case, how can lightning and thunder pre-exist, whereas phenomena associated with them, such as rain, snow, and hail, do not pre-exist, but are produced?

In Chapter 1 of Book C, Aristotle moves into whirlwinds, violent winds, and tornadoes. As he has stated many times, both moist and dry exhalations exist inside clouds. When the dry exhalation is ejected from the clouds in small, scattered quantities, it gives rise to thunder and lightning. If it is denser and rapid then it produces violent wind (windstorm).[15] If moist exhalation is ejected then we get a lot of rain.

When clouds interact (collide), it may happen that the wind inside a cloud is forced to go from a wider space to a narrower space. In this case, the front part of the wind is deflected because of the resistance of the narrowness or because of an opposing current. Then a Cyclone (wind moving in a circle) and a vortex may be created. This is how whirls and whirlwinds emerge. Whirls on the ground are created by the same process. But, in the case of whirlwinds, which are produced from the continuous ejection of dry exhalation from the cloud, they are accompanied by the body of the cloud. Because the wind cannot be freed from the cloud, not only rotates but it descends toward the ground. This is called tornado.[16] When a tornado moves in a straight line, its rotating motion sweeps away, overturns, and violently lifts everything on its path

---

[13] Similar to the sound a branding object makes when it is sunk into water, the hissing noise is 'fire' being extinguished in contact with the moistness of the cloud.
[14] Empedocles thought that light has a finite speed and thus it can be entrapped inside clouds. Aristotle does not accept this because according to him light and vision are instantaneous. More on that in our discussion on optics in Book C.
[15] Some translations call this hurricane.
[16] In the actual text, Aristotle uses the word τυφώνας, which nowadays is used for typhoons or hurricanes. However, Aristotle clearly describes a tornado here. Bearing in mind that hurricanes are extremely rare in the Mediterranean it is most likely that Aristotle never experienced one as we know it today.

higher. Tornadoes are not formed when the wind is a north wind, neither windstorms form when it snows, because they are wind and wind is dry/warm exhalation. When the wind is north or when it is snowing, it means that it is cold, which ceases the dry/warm exhalation.

Here are the main points on stormy weather which Aristotle makes in the last chapter of Book B and first chapter of Book C:

1. Dry exhalation trapped inside clouds is ejected when clouds collide creating thunder.
2. This ejected wind is usually inflamed and burns with a thin and faint fire. This is lightning.
3. Thunder occurs first. Lightning follows.
4. Violent winds (windstorms) are produced when the dry exhalation is dense and rapid.
5. Whirlwinds and tornadoes are formed when wind is forced to pass from a wider space to a narrower space.

Meteorology now, part 3

Stormy weather is associated with heavy precipitation and strong winds. In general, clouds capable of delivering such conditions are so-called cumulonimbus clouds. When these clouds are accompanied by lightning and thunder they are called thunderstorms. A thunderstorm is not just a cumulonimbus. It is a huge cumulonimbus, which grows to be 40,000 feet tall. A thunderstorm reaches the top of the troposphere, and often has enough momentum to penetrate the lower levels of the stratosphere. A thunderstorm develops when the rising motion is very strong, in other words, either when the heating at the surface is very strong, or when a warm and humid air mass collides with a much colder air mass. In the former case, the thunderstorm occurs within a single warm air mass during the warm humid summers. In the latter case, it forms along the cold front where the warm air is lifted by the colder air.

In either case, there are three distinct stages in the lifetime of a thunderstorm. The first stage is called the cumulus stage. In this stage cumulus clouds (small puffy clouds that look like cotton balls) grow and merge, slowly building a cumulonimbus cloud. Rising motions dominate the developing cloud. Once precipitation develops it begins to fall. The effect of falling precipitation is the creation of downdrafts, which oppose the rising motions inside the cloud. This is now the second stage, the mature stage. With precipitation continuing to develop, downdrafts begin to dominate the cloud.

Eventually the downdrafts cut off all rising motions, and the cloud enters its third stage called the final or dissipation stage. Since only sinking motions are inside the cloud, the sinking air warms, its relative humidity drops and because of that the cloud begins to evaporate.

A thunderstorm is considered to be severe when one or more of the following criteria are satisfied: hail with a diameter of three quarters of an inch or greater, wind gusts of 58 miles or stronger, and/or a tornado. The most severe of thunderstorms are the so-called supercell thunderstorms. They are huge single cell thunderstorms, which produce hail that can grow to the size of a grapefruit, 100 miles per hour wind, and most of the observed tornadoes. They are different than other thunderstorms because they rotate. Inside these storms the rising motion is rotating. The rotation is achieved because the wind in the horizontal (the atmospheric flow in which the rising motions are embedded in) changes in strength and direction with height in a more dramatic fashion than in typical thunderstorms. This change of the wind speed with height is called vertical wind shear. In environments of strong vertical wind shear, the rising motion breaks from its vertical path and begins to spin. As we will see in detail later, this is the first ingredient in tornado formation.

A typical thunderstorm can deliver about four million tons of water! But, this is not the only problem with thunderstorms. Even when they do not produce tornadoes, they produce wind damage, hail, and, as their name suggests, they generate lightning and thunder. Lightning has been marveled and feared alike since the beginning of time. Naturally, its grandeur and destructiveness were often identified with some deity. For the ancient Greeks it was Zeus, the King of Gods, who used it to punish those who disobeyed or displeased him. In the Viking mythology it was the product of Thor's hammer striking an anvil as he travelled through the clouds. Native Americans thought that lightning was the result of a mystical bird's flashing feathers. As it moved through the clouds, the clapping of its wings produced thunder. In the East, Buddha was depicted as carrying thunderbolts.

Observations have documented that inside the cloud the higher levels become positively charged and the lower part becomes negatively charged (Figure 21). As ice crystals high within a thunderstorm cloud flow up and down in the turbulent air, they crash into each other. As they crash past each other, small negatively charged particles called electrons are knocked off from some crystals and added to other crystals. This separates the positive (+) and negative (-) charges of the cloud. The top of the cloud becomes positively charged, while the base of the cloud becomes negatively charged. As the charged cloud moves over ground, friction causes a positive charge to accumulate at the ground. This positive charge follows the outline of the terrain under the cloud. Thus, if a tree or a building is present, the positive charge will form around them.

Figure 21: Charge distribution with a thunderstorm.

Nature does not like to be charged either positively or negatively. It much prefers the neutral state. Because of that, given the above set up, a discharge will take place between the bottom of the cloud and the underlying surface. Electrons surge from the cloud base toward the ground in a series of steps called the *step leader*. The step leader is rather faint and forked in structure. The complex geometry of the step leader is due to the fact that during the discharge, the charges take a path of *least resistance*. Since the atmosphere is not homogeneous in any respect, this path is often complicated. It is like trying to build a road between points A and B in a topographically complex, ragged region. The path of least effort (more economical) may not be the straight line connecting A and B. You may think of this as a process in which the channels through which the positive charge on the ground and the surging negative charge will communicate opens. Once the step leader reaches the level close to the ground, the positive charges rush upward to meet the step leader and neutralise each other via a powerful stroke. This is the flash we actually see as lightning. Discharges between clouds and the ground are called cloud-to-ground lightning. Cloud-to-cloud discharges between negatively and positively charged regions of two clouds are also possible. The leader-stroke process is often repeated within the same channel. The subsequent leader is called the dart leader; it proceeds downward easier now that the channel has been opened. The return stroke from the ground is weaker because most of the positive charge has been neutralised. A lightning flash may have four or more dart leaders separated by intervals of about four hundredths of a second. Because our eyes cannot react in such short times it appears that the flash flickers. Because of the principle of least resistance, chances are that the closest positive charge to the approaching step leader will jump into the opening channel first. As a result, lightning tends to strike the tallest objects in the area.

Lightning is a discharge of electricity. A single stroke of lightning can heat the air around it to 30,000°C (54,000°F).[17] This is more than five times greater than the temperature on the surface of the Sun! This extreme heat causes the air to expand explosively fast. The expansion creates a shock wave that turns into a booming sound wave, known as thunder. Thus, while thunder and lightning occur at roughly the same time, strictly speaking *lightning happens first*. The fact that we see the flash of lightning before we hear the thunder, is because the light travels much faster than sound (300,000 km/sec versus 0.34 km/sec).

The rotating rising motion inside a supercell thunderstorm is called the *mesocyclone* (Figure 22 top left). This mesocyclone may be stretched in the vertical. Exactly how this stretching happens is not clear. Evidence suggests that the lower part of the rising motion inside the cloud is slowed down by other movement in the supercell thunderstorm. As a result, the upper part is now rising faster and the entire column stretches in the vertical and shrinks in the horizontal. Another possibility is that the motion accelerates due to extra heating from the latent heat of condensation inside the cloud. Be this as it may, from this point on, a law of physics called the conservation of angular momentum is going to take over and make a tornado. The angular momentum refers to an object that rotates. It is equal to the mass of the object times its rate of rotation times the radius of the object.

Angular momentum=mass x radius x rate of rotation

According to this law, the product on the right-hand side of the equation remains always the same. For this to happen, if one variable goes up the other two have to change so that the product remains unchanged. For example, if the mass of the object remains constant, an increase in the rate of rotation must cause a decrease in the radius and vice-versa. Similarly, a decrease in the rate of rotation must be accompanied by an increase in the radius and vice-versa. It follows that, if the mesocyclone's radius decreases because of the stretching in the vertical, the rate of rotation should increase. This law is put into practice by skaters. When skaters want to spin faster,

---

[17] The atmosphere does not get hot during a thunderstorm because lightning is extremely thin, and heats very small amount of air, which then uses gained heat to rapidly expand.

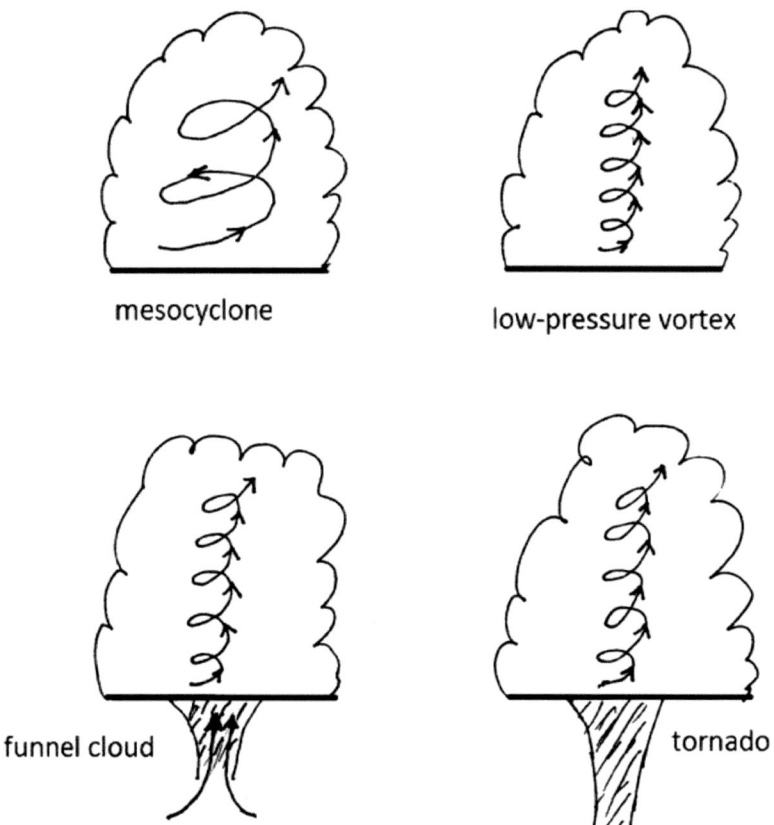

Figure 22: Stages in the formation of a tornado.

they begin spinning with hands and legs extended. Because the hands and legs are extended, the radius of the skaters is large. Drawing in their hands toward the axis of rotation decreases the radius and they spin faster. When this happens in the cloud, a very fast rotating *low-pressure vortex* happens (Figure 22, top right). This low-pressure vortex acts like a powerful vacuum that begins to suck the air from below the cloud base. As this air rushes in this vortex, it cools rapidly and condenses into a visible cloud (Figure 22, bottom left). This is the *funnel cloud*, which appears to descent as the rushing air keeps on condensing. Once the funnel cloud reaches the surface it is called a *tornado* (Figure 22, bottom right). When a tornado touches the ground it picks up dirt or whatever is on its path. However, contact with the ground cuts off the ingredient necessary for its formation and sustainability (the rising air) and the tornado will soon die out.

While there is a good chance that a supercell and a mesocyclone will produce tornadoes, not all of them do. All the conditions have to be right for a storm to produce a tornado. Also, it is possible to get a tornado without a mesocyclone. In this case no tubes of spiraling air are formed because of vertical wind shear. The differences in wind direction and speed at some level in the horizontal may break the horizontal flow into more or less planar vortices, which may then develop when interacting with updrafts. When this happens the tornadoes are weaker. Such tornadoes are called landspouts because of their similarity to waterspouts, another tornado vortex which is sometimes observed over bodies of water. Other tornado-like vortices include the *dust devils*. Dust devils are associated with dry convection (rising dry air). They form over hot surfaces, typically in deserts, and are rather weak.

Key points of meteorology now, part 3

1. A thunderstorm is associated with thunder and lightning. Thunderstorms produce heavy rain and/or hail, and may give birth to a tornado.
2. Inside the developing cloud, a separation of charges takes place. This separation makes the top of the cloud positively charged, and the bottom of cloud negatively charged.
3. As the storm moves over a surface, friction between the surface and the bottom of the cloud causes the surface to become positively charged.
4. Lightning ensues as a result of the discharge of those opposite charged fields.
5. Lightning is very hot and causes nearby air to expand explosively fast. This expansion creates a shock wave, which turns into a booming sound, the thunder.
6. The rising motion inside the thunderstorm may under certain conditions rotate. This rising spinning motion may then be stretched in the vertical and shrink in the horizontal. If this happens, according to the conservation of angular momentum law, a low-pressure vortex develops, which leads to the formation of a tornado.

Back to Aristotle's *Meteorologica*

In Book B and chapter 1 of Book C, Aristotle tried to explain the causes of rain, winds, and other weather phenomena that affect our lives, such as thunder, lightning, severe storms, thunderstorms, etc. As we discussed previously, Aristotle was not aware of the concept of atmospheric pressure and its importance in atmospheric processes. Aristotle's adaptation of dry and moist exhalations as his 'axioms' and his view that dry exhalation causes wind, did not lead him to a correct description of the genesis of winds. Similarly, his exhalations, while providing him with a 'logical' procedure for the description of thunder, lightning, and stormy weather, failed to provide the correct understanding of these phenomena. It can be argued that his explanations of thunder and lightning may have been rooted in his fundamental axiom on relationships between the four principles and four elements, according to which the fire is caused by the combination of dry and hot (recall Figure 2). While we find his analogy of the 'laugh of Hephaestus' to lightning and thunder rather amusing, the major points made in chapter 9 of Book B and chapter 1 of Book C:

1. Dry exhalation trapped inside clouds is ejected when clouds collide creating thunder.
2. This ejected wind is usually inflamed and burns with a thin and faint fire. This is lightning
3. Thunder occurs first. Lightning follows
4. Violent winds (windstorms) are produced when the dry exhalation is dense and rapid
5. Whirlwinds and tornadoes are formed when wind is forced to pass from a wider space to a narrower space,

are not correct except for the last one. In his discussion on tornadoes, Aristotle manages to describe a key element, which today is known as the conservation of angular momentum law, which is correct.

BOOK C FROM ΜΕΤΕΩΡΟΛΟΓΙΚΑ

Aristotle's optics

Preparatory introduction

In Book B we already discussed the first chapter of Book C. Thus we are starting discussion of Book C with the second chapter, which is an introduction of the four atmospheric optical phenomena, namely, halos, rainbows, sundogs,[1] and light pillars. In this chapter Aristotle, largely based on observations, writes about their characteristics, after he states that all four phenomena are the result of the same cause. Here are his main points, which will be discussed in more details in following sections:

Halos often appear as a complete circle either around the Sun or the moon or, sometimes, around bright stars.[2] They form by night as well as by day (close to midday or afternoon, and more rarely around sunrise or sunset).

Rainbows, on the other hand:

1. Never form a complete circle, nor any part that is greater than a semicircle.
2. During sunrise and sunset, their arch is greater and their height is smaller. As the Sun rises higher, the height increases and the arch decreases. After the autumn equinox, when the days get shorter, rainbows form at any time of the day, while in summer they do not form at noon.
3. There are never more than two rainbows at the same time. When there is a double rainbow, each has three colours,[3] but in the outer (secondary) rainbow the colours are fainter, and their position is reversed, compared to the inner

---

[1] A sundog is formally called parhelion in meteorology from the ancient Greek παρηλιον, which means on the side of the Sun.
[2] Aristotle refers to Venus and Jupiter.
[3] The colours of the rainbow are seven: red, orange, yellow, green, blue, indigo, and violet. Aristotle unites red, orange, and yellow to what he calls οινικουν (red), and blue, indigo, and violet to what he calls αλουγρον (product of the sea; or purple).

(primary) rainbow. In the primary rainbow, the first (top) band is wider and red, whereas in the secondary red is the narrowest and third (bottom) band. In this way, the red bands are next to each other. The space between red and green often appears as yellowish.
4. These three colours are the only colours painters cannot reproduce. Painters make their colours by mixing other colours, but red, green, and purple cannot be manufactured by mixing other colours.
5. Rainbows may also form at night under the influence of the moon. It is a rare phenomenon and it often goes unnoticed, because, in the dark, colours are not easily detected. For this to happen, the moon should be full, and rising or setting.[4]

Sundogs and light pillars are always formed on the sides of the Sun, not above or below it, and usually when the Sun is rising or setting (mostly at setting). Very rarely, they may form when the Sun is overhead.[5] They don't form during the night.

According to Aristotle, the common process behind the formation of all four phenomena is reflection. The differences are due to the ways the reflection interacts with the surface, and due to properties of the source, which can be either the Sun or some other bright object. Aristotle ends this chapter as follows: 'Given what we know from the theory of optics, and from the fact that in some mirrors we can see our reflection, whereas in others we see only colours, we must conclude that, in the same way a ray is reflected from water, rays would be reflected from the air and from any objects with a smooth surface. The type of mirrors where we see only colours, are those of very small size so that they are indivisible by sense.[6] In such mirrors, it is indeed impossible for a shape to appear (because otherwise the mirror would be considered divisible, since every shape is considered, in general, to be, at the same time, a shape and divisible). But since something must be reflected, and it is impossible to be the shape, then what remains to be seen is the colour. With respect to the colour of bright objects, sometimes it appears bright, whereas sometimes owing to the admixture of the colour of the mirror or to weakness of sight, it appears as some other colour.'

Aristotle's general theory of colour

Aristotle's central point in his colour theory is that light and darkness, or white and black, when mixed, produce other colours. Red contains more white and less black

---

[4] Aristotle mentions that because of these conditions, a moon rainbow was observed only twice in 50 years.
[5] Here Aristotle refers to an alleged incident in the area of Bosporus, when two sundogs appeared on each side of the Sun and followed it all day until sunset.
[6] For Aristotle these mirrors come from condensation of water vapour, i.e. cloud droplets or small raindrops. He never mentions ice crystals.

than other colours; green has more black and less white than red, and violet possesses still more black and less white. This is equivalent to saying that as light gets gradually weaker or darker, the colours are produced in succession. Or, expressing this view in terms of the theory of visual rays, colours are produced by the weakening of sight.

Aristotle considers that the rainbow contains three colours: red, green, and violet (purple). He sometimes also mentions yellow (or orange), but this colour is not to be considered in the same class as the three mentioned above; it is due only to contrast. Red, green, and violet are the primary, the basic colours: they are real colours. This tradition of considering only three colours as basic and real continued for a long time and remained unchallenged in the Aristotelian form until the beginning of the 14th century. The arguments Aristotle used to justify those three colours can be classified as: a) mystical or philosophical; b) arithmetical; c) experimental.

a) When sight becomes less white, i.e., when sight is weakened and approaches black (i.e. departs from white) the colours are produced. As he states 'When sight is relatively strong the change is to red; the next stage is green, and a further degree of weakness gives violet. No further change is visible, but three completes the series of colours, and the change into the rest is imperceptible to sense. Hence the rainbow appears in three colours.' The following quotation from his book *On the Heavens* will make clear how three completes the series of most things: 'Now a continuum is that which is divisible into parts always capable of subdivision, and a body is that which is every way divisible. A magnitude if divisible one way is a line, if two ways a surface, and if three a body. Beyond these there is no other magnitude, because the three dimensions are all that there is, and that which is divisible in three directions is divisible in all. For as the Pythagoreans say, the world and all that is in it is determined by number three, since beginning, middle, and end give the number of an 'all', and the number they give is a triad. And so, having taken these three from nature, we make further use of number three in the worship of the gods. Of two things, or men, we say 'both', but not 'all ': three is the first number to which the term 'all' has been appropriated. Therefore, since 'every' and 'all' and 'complete' do not differ from one another in respect to form, but only, if at all, in their matter and in that to which they are applied, body alone among magnitudes can be complete. For it alone is determined by the three dimensions, that is, is an 'all'.'

b) In *Sense and Sensibilia* Aristotle states: 'Firstly, white and black may be juxtaposed in such a way that by the minuteness of the division of its parts each is invisible while their product is visible and thus colour may be produced. This product can appear neither white nor black, but, since it must have some colour and can have neither of the above two, it must be a sort of compound and a fresh kind of tint. In this way then, we may conceive that numbers of colours over and above black and white, may be produced and that their multiplicity is due to differences in the proportion of their composition. The juxtaposition may be in the proportion of 3 to 2, or 3 to 4, or according to other ratios. Others again may be compounded in no commensurate proportion, with the excess of one element and deficiency of the other which are incommensurable, and colour may, indeed, be analogous to harmonies. Thus, those compounded according to the simplest proportions, exactly as is the case in harmonies, will appear to be the most pleasant colours, e.g. purple, crimson, and a few similar species (it is exactly the same reason that causes harmonies to be few in number). Mixtures not in a calculable ratio will constitute other colours. Or again, all tints may show a calculable proportion between their elements, but in some the scheme of composition may be regular in others not, while then those of the latter class are themselves impure, which may be due to an absence of calculable proportion in their composition.'

c) Back to *Meteorologica* Book C: 'In the inner rainbow the first and largest band is red: in the outer rainbow the band that is nearest to this one and the smallest is of the same colour. The other bands correspond on the same principle. These are almost the same colours which the painters cannot manufacture. There are colours which they create by mixing, but no mixing will give red, green, or purple. These are the colours of the rainbow, though between the red and the green an orange colour is often seen'. '... if the principles we laid down about the appearance of colours are true, the rainbow necessarily has three colours, and these three and no others. The appearance of yellow is due to contrast, for the red is whitened by its juxtaposition with green. We can see this from the fact that the rainbow is purest when the cloud is blackest; and then the red shows most yellow. Therefore, yellow is some intermediate colour between red and green. Red appears lighter by contrast with the blackness of the cloud around... The moon rainbow affords the best instance of this colour contrast. It looks quite white: this is because it appears on the dark cloud and at night'.

It is implicit in the above statement that yellow is nearer to white than to red. Here are some more statements by Aristotle concerning colour contrast: '...just as fire is intensified by added fire, black next to black makes that which is in some degree white to appear whiter; and this happens with red. Bright dyes show the effect of contrast as well. In woven and broidered stuffs, the appearance of colours is profoundly affected by the juxtaposition with each other, and also by difference of illumination. Thus, embroiderers say that they often make mistakes in their colours when they work by lamplight, and use the wrong ones (dyes)'.

We have mentioned that according to Aristotle colours are produced by the mixture of white and black or by the weakening of sight or light. As will be seen from the quotations to follow, this weakening occurs in three ways: a) reflection; b) distance; c) degree of opacity or darkness of the reflector. Aristotle mentions a number of phenomena or observed facts which he considers as related to the rainbow. Most of these are used mainly to justify his general colour theory.

> 'Rainbows are seen around the lamps, in winter, generally with southerly winds. Persons whose eyes are moist see them most clearly because their sight is weak and easily reflected. The formation of colours in this sort of bow is due to the dampness of the air and to the soot created by the flame, that mixes with air and makes it a mirror, and the blackness that the mirror derives from the smoky nature of the soot.'

> 'We know how red the flame of green wood is: this is because a lot of smoke is mixed with the bright white firelight: thus also the sun appears red through smoke and mist.'

> 'Whenever the sight is strained and fails, distant objects and objects seen in a mirror look darker, smaller, and smoother, and the reflection of clouds in water is darker than the clouds themselves. This latter is clearly the case: reflection diminishes the sight that reaches them. It makes no difference whether the change is in the object seen or in the sight, the result being in either case the same... Clearly, then, when sight is reflected it is weakened, and as it makes dark look darker, so it makes white look less white changing it and bringing it nearer to black.'

The halo

As we was already mentioned, according to Aristotle, halos are formed when the visual ray is reflected by a mist made out of small parts, which he calls mirrors. He provides the following figure (Figure 23) to explain why halos form circles (around the Sun or the moon).

Aristotle considers visual rays[7] OA, OB, and OD. By construction, in three dimensions we have two right cones (AOB and ASB) whose base is the halo (in blue). Then, OA=OD=OB, and AS=DS=BS. Given that, and because triangles OAS, ODS, and OBS have the same

---

[7] There is a slight confusion/inconsistency in Aristotle's writings regarding transmission of light. In his books *Sense and Sensibilia* and *On the Soul*, he treats light as being transmitted from the source to the eye. By visual rays he means light produced by the eye. This inconsistency may be reconciled by considering that 'propagation' of light as is used in those two books is not what the word propagation means today. In the Aristotelian conception of propagation of light, no velocity is involved. It is instantaneous. As he states 'Light is due to the presence of something but is a motion'. This will make light a unique property of the Universe. Quite an insight, given that the speed of light of 300,000 kilometers per second is a fundamental constant and the upper limit of any motion in the Universe. For this reason, in this book we don't discriminate between these two views.

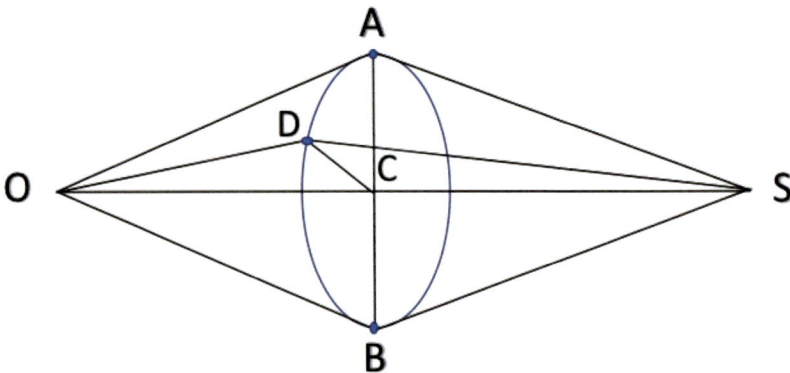

Figure 23: Aristotle's explanation of why halos are circular. O stands for observer and S for Sun (or moon).

base (OS), they are equal. It follows that the heights AC, BC, and DC are equal and they meet each other at point C, the centre of the circle (halo) ADB. Aristotle states that the mirrors must be thought of as being continuous: each of them is too small to be visible, but their contiguity makes the whole seem as one. What we see inside the halo is white colour[8] (the Sun), which is seen as a continuous circle, appearing successively in each of the mirrors and indivisible to sense. The ring appears darker because of the contrast with the white colour. Aristotle has been correctly criticised here for 'begging the question', meaning that he assumes what he has to prove. He a priori assumes that points A, B, and D are on a circle.

While Aristotle was aware of refraction, (for example he states in chapter 4 ... 'white light through a dark medium appears red'), he does not differentiate between reflection and refraction. His whole emphasis is on reflection. In the upper atmosphere, cirrus clouds are very thin and are made of tiny ice crystals. Due to their hexagonal structure, when light goes through them, it exits refracted by $22°$ (Figure 24).

Now let us consider Figure 25. Because the cloud is very thin, the ice crystals can be thought as existing on a plane. The vertical grey line is part of this plane. Point A is the eye of the observer. Because of the $22°$ refraction, the observer intercepts only the solar rays indicated by the orange colour (points C and D). All ice crystals in the cloud deviate the light similarly, but only the ones from the specific ring at $22°$ contribute to the effect for an observer at a set distance. In the figure, the other rays (black) are refracted but they don't reach the eye. If AB is the vertical line from A to the grey line, then triangles ABC and ABD are equal because they have a common side (AB) and all angles equal. Therefore, BC=BD. By rotating ABC around side AB, we can extend this two-dimensional

---

[8] Recall that according to Aristotle the mirrors are too small to reflect shape; they reflect colour.

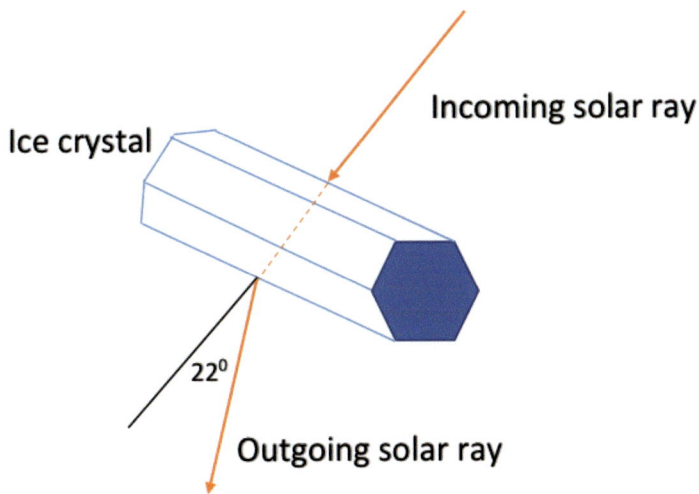

Figure 24: Refraction of light by an ice crystal.

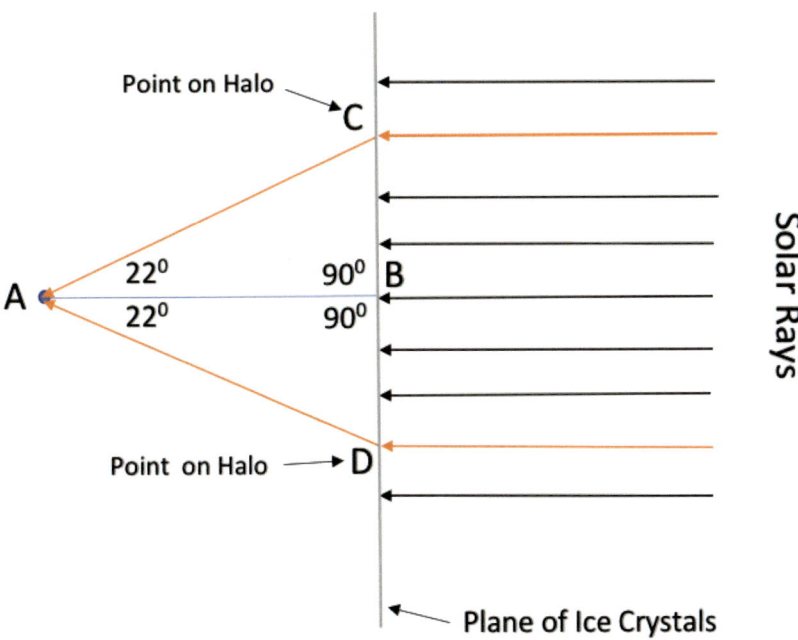

Figure 25: Illustration of halo formation. See text for details.

figure into a three-dimensional figure in which all Cs and Ds will be equidistant from point B and on a circle, which will be the base of right cone ACD (the halo). As no light is refracted at angles smaller than 22°, the sky is darker inside the halo.

Rainbow

In chapters 4 and 5 Aristotle deals with rainbows. Most of the points made in chapter 4 have been discussed within Aristotle's theory of colour, such as his arguments on the basic principle behind the formation of colours, the three colour rule, the structure of the double rainbow, and some other observations related to rainbows. In chapter 5 Aristotle presents an extremely complicated geometric treatment to prove why rainbows are circular. We will try to go through his logic step by step using simple examples and diagrams.

His general set up is that the observer is at the centre of a sphere, and between the Sun and the cloud. Both the Sun and the cloud are on the surface of the sphere. The rays from the Sun are reflected by the cloud and reach the eye of the observer who sees the rainbow. The Sun and the rainbow are represented as equidistant from the observer. Thus, the Sun is treated as it is at a finite distance.

We have already discussed that because of the existence of the tiny mirrors (raindrops), it is indeed impossible for a shape to appear. But since something must be reflected, and it is impossible to be the shape, then what remains to be seen is the colour. This settles why the returning reflection is not the image of the Sun. To explain why the image has the shape of an arc we produced Figure 26. It shows the observer (O) at the centre of a sphere and the Sun on the surface of the sphere indicated by the letter S. Imagine any plane passing through the point of observation and the Sun, and consider the intersection of this plane with the surface of the mirror (the cloud), which exists in front of the observer. This intersection is a line and a radial section of the rainbow is a small part on this line. And it is an observed fact that this section on the mirror is functional in transmitting an image to the observer. In the figure, this section is indicated by M, it should be thought of it as a line of three colours (red, green, violet). If the location of this section is determined by the relative positions of Sun-observer-bow (also referred to as 'geometrical uniformity'), then the locus[9] of all Ms would give the rainbow. Therefore, the locus of all Ms, which is the totality of all Ms on all the planes which pass through the sun and the observer, and which are normal to the surface of the mirror (because the plane passes through the centre of the sphere), will give the shape of the rainbow, and this locus is a circular belt. This locus is obtained by rotating SOM around SO as axis. The result is a right cone the base of which is the

---

[9] The locus is a curve, or any other geometric object, formed by all points satisfying a particular condition.

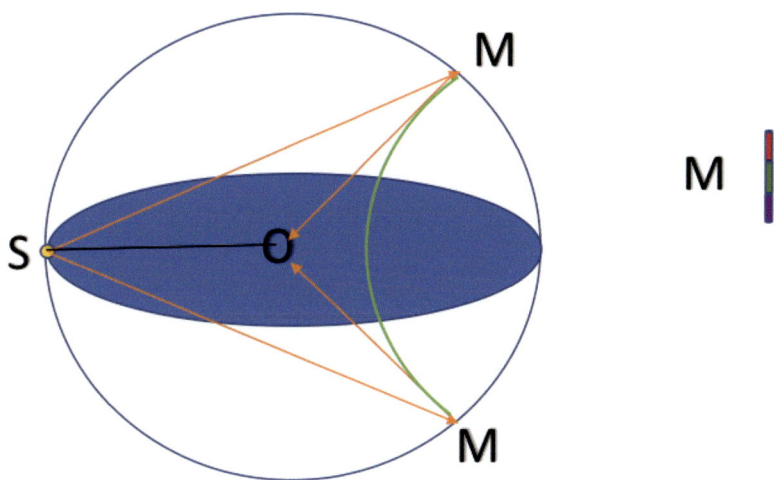

Figure 26: Aristotle's set up and geometric illustration of rainbow production. S is the Sun, O is the observer at the centre of a sphere. The cloud is in front of the observer.

locus. Thus, the rainbow is a circle (green curve). The fact that we only see part of this circle is because visualising the whole circle is limited by the horizon.[10]

We will come back to Aristotle, but first let us discuss what we now know about the rainbow today.

When light strikes a raindrop, it is not just reflected. It gets refracted as it enters the raindrop, then it reflects internally off the other side of the drop, and emerges again from the opposite side, again with refraction. Refraction is the bending of light when it enters a medium with a different density from the one through which it has been travelling. Much like in the case of light going through a prism, this process results in the splitting of light into its components: the spectrum of the visible light (Figure 27). As we all know light travels in forms of waves and each wave is characterised by its wavelength (the distance between two crest of the wave). Red has the highest wavelength and violet the lowest. Because of that, the angle between incoming solar ray and outgoing red light is 42⁰, whereas that of solar ray and outgoing violet is 40⁰ (Figure 28).[11]

---

[10] When high in the atmosphere (say in an airplane) the full circle can actually be seen.
[11] Note here that the same should happen with the refraction described in the formation of halos. However, because the refraction angle when ice crystals refract is much smaller, the effect of splitting is not as pronounced.

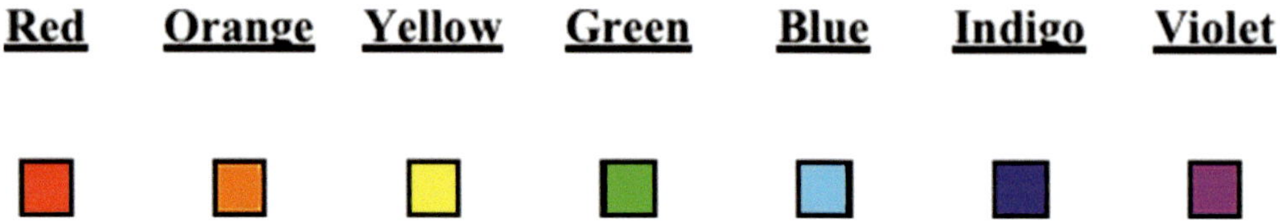

Figure 27: The spectrum of the visible light.

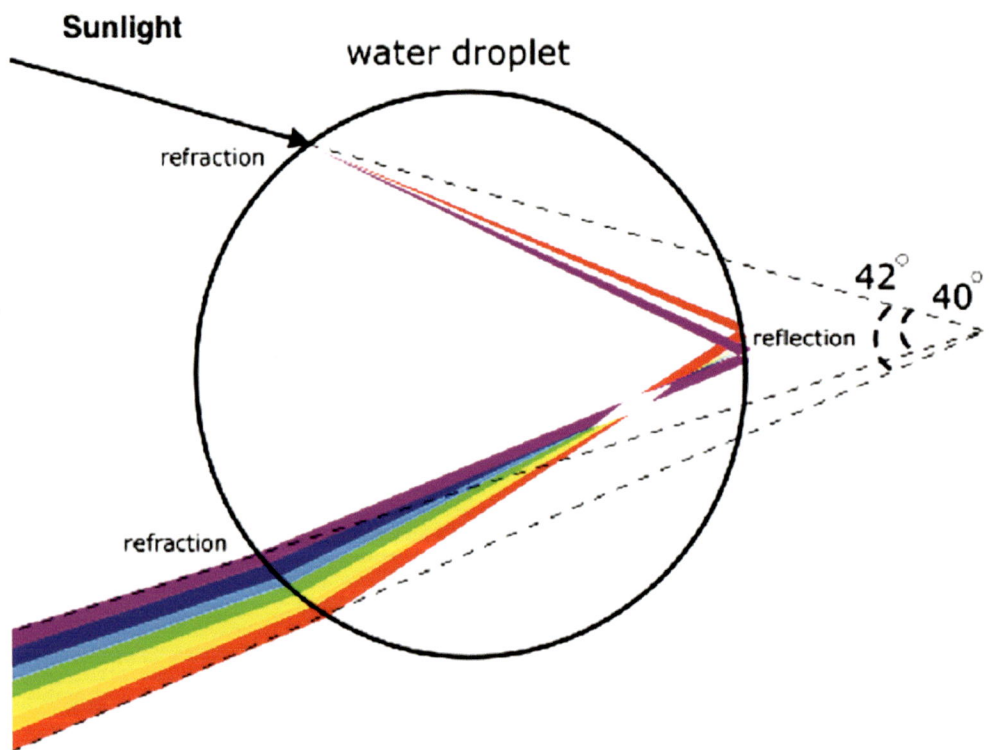

Figure 28: As light from the Sun enters a raindrop it undergoes refraction, reflection and refraction again, a process that splits it into its colour components.

Where in the sky does one need to look to see the light returning from the raindrops? First, we define a direction in space exactly opposite the sun, called the *antisolar point*. The antisolar point is marked by the shadow of the head (Figure 29). To see the light coming back from the raindrops, one has to look 42° away from the antisolar point. Of course, the region of the sky 42° away from that point is not just one direction but a whole collection of directions. As we see in Figure 29, the angle between the incoming sunlight and the red light (broken lines corresponding to two raindrops), is 42° for all points forming a circle around the antisolar point. The set of all the raindrops that have the same angle between the observer, the drop, and the sun, lie on a cone pointing at the sun with the observer at the tip. The base of the cone forms a circle at that angle with the antisolar point at its centre. Because of the different refraction, the rainbow circle is not 42 degrees for all colours. The circle is smaller for blue rays

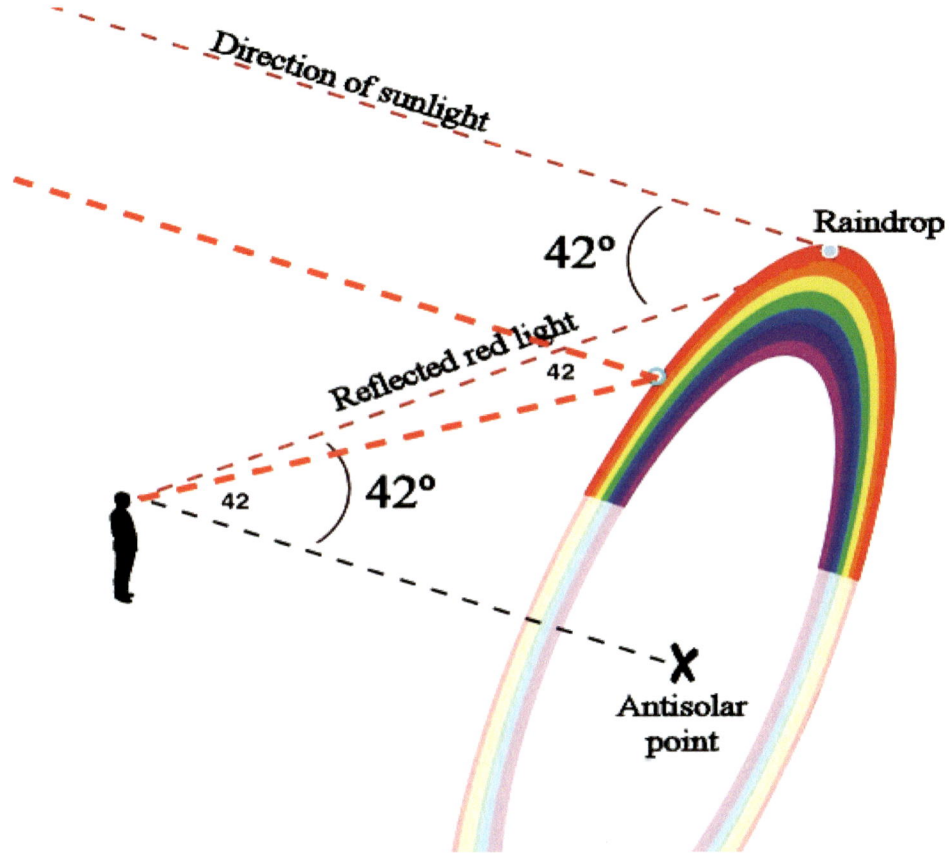

Figure 29: Formation of rainbow.

than it is for red rays, so the primary rainbow is violet on the inside and red on the outside, with the intermediate colours of the spectrum spread out in between.

It is also possible for some of the light rays entering the drop to undergo not one but two internal reflections before they emerge. If one looks at the whole collection of such rays, a concentration of light coming back at an angle of 51° can be seen. These rays produce a secondary rainbow, also centred on the antisolar point, having an angular radius of 51°; the secondary rainbow therefore appears outside of the primary bow. The sequence of colours is reversed in the secondary bow (see Figure 30).

Going back to Aristotle, despite him not considering refraction, both in his explanation of the halo and of the rainbow he knows that a uniformity is involved in the reflection of light: 'Since the reflection takes place in the same way from every point… we must accept from the theory of optics the fact that sight is reflected from air and any object with a smooth surface just as it is from water…' He knows that there is a law according to which reflection takes place, but he does not know, and also, perhaps he does not feel the need of knowing the specific form of the law itself. He knows that if light (or sight) K proceeding from K is to be reflected from the surface Q (Figure 31) to the point H, there is a unique point M from which this reflection must take place; and that any other path such as KPH will not do. But he does not know the correct way of determining the point M. According to Aristotle if M is a point where reflection from K to H can occur, then this reflection is determined by the ratio KM/MH. The point

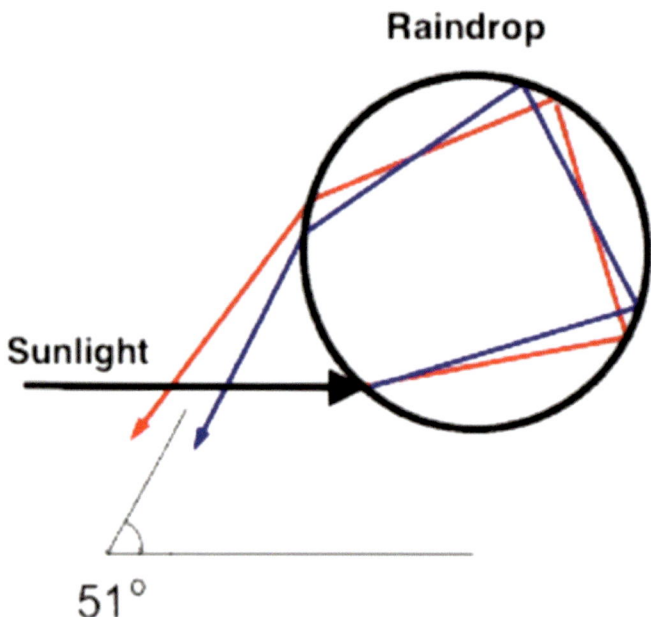

Figure 30: Formation of the secondary rainbow.

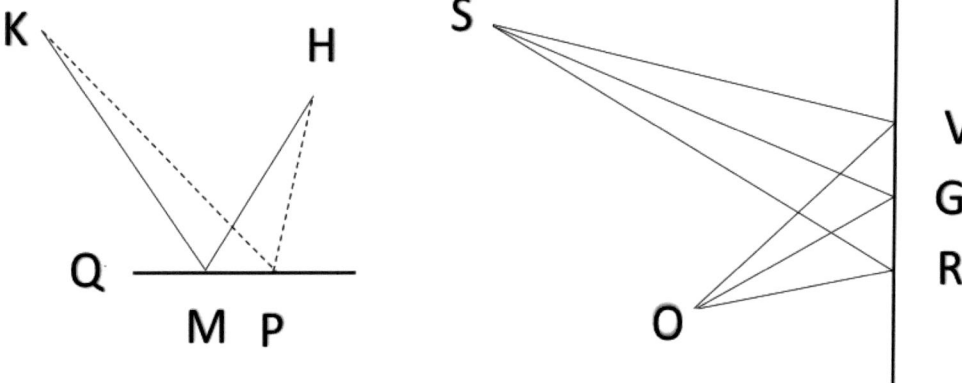

Figure 31: Aristotle's law of reflection. Reflection to H from K can only occur from point M.

Figure 32: Sequence of colours in the secondary rainbow.

P cannot reflect light from K to H, because KP/PH ≠ KM/MH. Besides mentioning the existence of a regularity in the reflection of light, Aristotle makes use of the regularity in accounting for the shapes of both the rainbow and the halo, and this regularity is the most essential point in the explanation of the shapes.

Given Aristotle's set up (Figure 26), clouds on a hemisphere resting on the circle of the horizon reflect sunlight to the observer where the angle (SMO) is equal. Aristotle does not say what the angle is, however, as we discussed above we know that 42^0 is a good choice.

Under these conditions Aristotle's explanation is perfectly legitimate. It is a geometrical scheme of approach, no doubt suggested by the geometrical uniformities already known to Aristotle between the position of the bow, and the locations of the sun and the observer. Interestingly, Aristotle's theory of rainbow formation has been applied to construct a method for real-time rendering of rainbows with very impressive results.[12] The serious difficulty in Aristotle's rainbow theory, shows itself in explaining the secondary bow.

In Figure 32, S is the Sun, O the observer, VGR is a radial cross section of the *secondary* rainbow (M in Figure 26 but with reverse order of the colours), and V, G, R are the locations of violet, green, and red, respectively. Aristotle colour theory says that, since

---

[12] Frisvad, J. R., Christensen, N. J. and Falster, P. 2007. The Aristotelian Rainbow: From Philosophy to Computer Graphics. In *Proceedings of GRAPHITE 2007: 5th International Conference on Computer Graphics and Interactive Techniques in Australasia and Southeast Asia ACM*. https://doi.org/10.1145/1321261.1321282.

OR<OG<OV, the reflection at R will be strongest, that at V will be the weakest; and the reflection at G will be stronger than that at V but weaker than that at R. He had already told us that as sight gets weaker the colour will depart from white and approach black. Hence red will appear at R, green at G and violet at V. All this is logical and consistent with his colour theory. But if this reasoning is to be adopted in explaining the order of the colours in the *primary* bow, the order of colours will not come out as they are observed, but in reverse order. Thus, in the case of the primary bow, Aristotle had to resort to an entirely different way of approach in order to explain the order of the colours. Here distances such as OR, OG, OV do not play any role. What is of importance is the total amount of reflection from each circular band. If we consider the sum total of the VGR reflected sight, this sum is a function of the size of the band. We know that colour is determined by the amount of brightness or darkness, or, in other words, is dependent on the amount of reflected sight. This amount is greatest in the case of the greatest arc. Therefore, the outer circle will be red. At the second and the inner circles green and violet will appear according to the same reasoning.

There were two major problems in the explanation of the rainbow: the shape and the colours. For the sake of accounting for the circular shape of the rainbow, Aristotle introduces certain simplifications such as the Sun and the cloud being equally distant from the observer, and both at finite distances. Some scholars have wondered whether Aristotle knew about Euclid's law of reflection or not.[13] He does use a principle of equal angles as the key to describe halos and rainbows by reflection off mist and clouds, respectively, but he does not specifically mention the Euclidean proposition. While for his explanation of the double rainbow it is necessary to employ two different processes, the Aristotelian explanation of both these points is very good within the limitations to which he was subjected; in both cases, one may say, he went as far as it was possible. All in all, considering the general level of scientific knowledge at the time, the Aristotelian explanation of the rainbow was a great contribution to scientific knowledge, and no real progress was made on the subject until the end of the 13th and the beginning of the 14th centuries.

Sun dogs and light pillars

The sun dog is a member of the family of halos, caused by the *refraction and scattering* of sunlight by plate-shaped hexagonal ice crystals in the atmosphere. These ice crystals are either suspended in high cirrus or cirrostratus clouds, or drifting in cold air at lower levels. Like in the formation of halos, the crystals act as prisms, bending the light rays passing through them by 22° (see Figure 24). As the crystals gently float downwards with their large hexagonal faces almost horizontal, sunlight is *refracted*

---

[13] Light travels in straight lines and reflects from a surface at the same angle at which it hit it.

horizontally, and sun dogs are seen as a pair of coloured patches around 22^0 to the left and right of the Sun, and at the same altitude above the horizon as the Sun. They can be seen anywhere in the world during any season, but are not always obvious or bright. Sun dogs are best seen and most noticeable when the Sun is near the horizon.

Light pillars are also caused by the interaction of light with ice crystals, and they, therefore, also belong to the family of halos. The crystals responsible for light pillars usually consist of flat, hexagonal plates, which tend to orient themselves more or less horizontally as they fall through the air. Each plate acts as a tiny mirror which *reflects* light sources which are directly above or below it, and the presence of plates at several altitudes causes the reflection to be elongated vertically into a column. The larger and more numerous the crystals, the more pronounced this effect becomes. They can be caused by sunlight, moonlight, and even street lights.

Thus, sun dog formation involves refraction and light pillar formation involves reflection. Figure 33 is a beautiful photo taken in Fargo, North Dakota where all three phenomena (halo, sun dogs, and a light pillar) coexist.

As we mentioned in the beginning, Aristotle considered that reflection of light is behind of the formation of all four optical phenomena. According to Aristotle, in both Sun dogs and light pillars, visual rays are reflected by tiny droplets (mirrors) from the condensation of water vapour. The difference is that in the case of sun dogs the mirrors are uniformly distributed and of the same density throughout, whereas in the case of light pillars, they are unevenly distributed. Since all three, halo, sun dogs, and light pillar may coexist, Aristotle is clearly not correct in assuming differences in air composition as the reason for the differences between sun dogs and light pillars.

Aurora Borealis

In Book A, chapter 5 Aristotle writes: 'Sometimes on a fine night we may see a variety of phenomena, which form in the sky: chasms, trenches, and blood-red colours. All those have the same cause. The upper air condenses into an inflammable condition and combustion gives them the appearance of a burning flame, sometimes that of moving torches and stars. It is not surprising that this same air when condensing would assume a variety of colours, because a weak light shining through a dense medium, when the air acts as a mirror, will cause all kinds of colours to appear, especially crimson and purple. These are the colours which generally appear when fire-colour and white are combined by superposition. Thus, on a hot day, upon rising and setting the stars look crimson when seen through a smoky medium. The air will also create colours by reflection when the mirror reflects colour but not shape (see previous discussion on Aristotle's general theory of colour)'.

Figure 33: Halo around the Sun, sun dogs on the right and left of the Sun, and a light pillar above the Sun. Courtesy of Wikimedia Commons. Attribution Gopherboy6956.

It is interesting to note that Aristotle's description of Aurora Borealis implies that he either heard of its the appearance and/or personally witnessed it. It is very rare to have Aurora effects seen at the latitude of Greece where he lived. In fact, there is only one description of such an appearance in the 1930s at the Observatory on the island of Spetses. Modern science says that the Aurora Borealis is the product of the excitement of atmospheric constituents in the upper atmosphere such as oxygen atoms and molecules, bombarded by fluxes of charged particles emitted at high speeds from the Sun. These are protons and other particles that the Sun emits at times of big magnetic storms happening at its surface. These storms are called solar flares and they carry with them a magnetic field from the Sun which collides with the magnetic field of the Earth. Indeed, the Earth is engulfed by its own magnetic field, which deflects the coming particles to the poles of the Earth. As they enter the Earth's magnetic field, they follow spiral trajectories which collide with atoms and molecules in the atmosphere. Most of the energy transported by the particles is dissipated at great levels in the atmosphere and the excited atoms and molecules in the atmosphere emit green, red, and purple light, which are typical lines when some molecules and atoms are excited. The shape of Aurora Borealis seen from space resembles that of a crown (see top of Figure 34), which is produced by the geometry of emissions by the excited particles in the atmosphere at high levels. The visible picture from the ground looks like a curtain (bottom Figure 34) that oscillates and moves around the north magnetic pole of the earth (thus creating the crown structure). We note here that the north magnetic pole does not coincide with the north pole of rotation of the Earth. It is now located off the coast of Greenland and in recent years is moving towards Siberia at a rate of around 34 miles per year.

Near to the North Pole, the phenomenon is called an *aurora borealis* or *northern lights*. Near to the South Pole, it is called an *aurora Australis or the southern lights*. It is interesting to note here Aristotle's considering 'combustion' to be a part of the formation of Aurora Borealis, which 'resembles' the excitement of particles in the atmosphere by the interaction of solar radiation and the magnetic field of the earth.

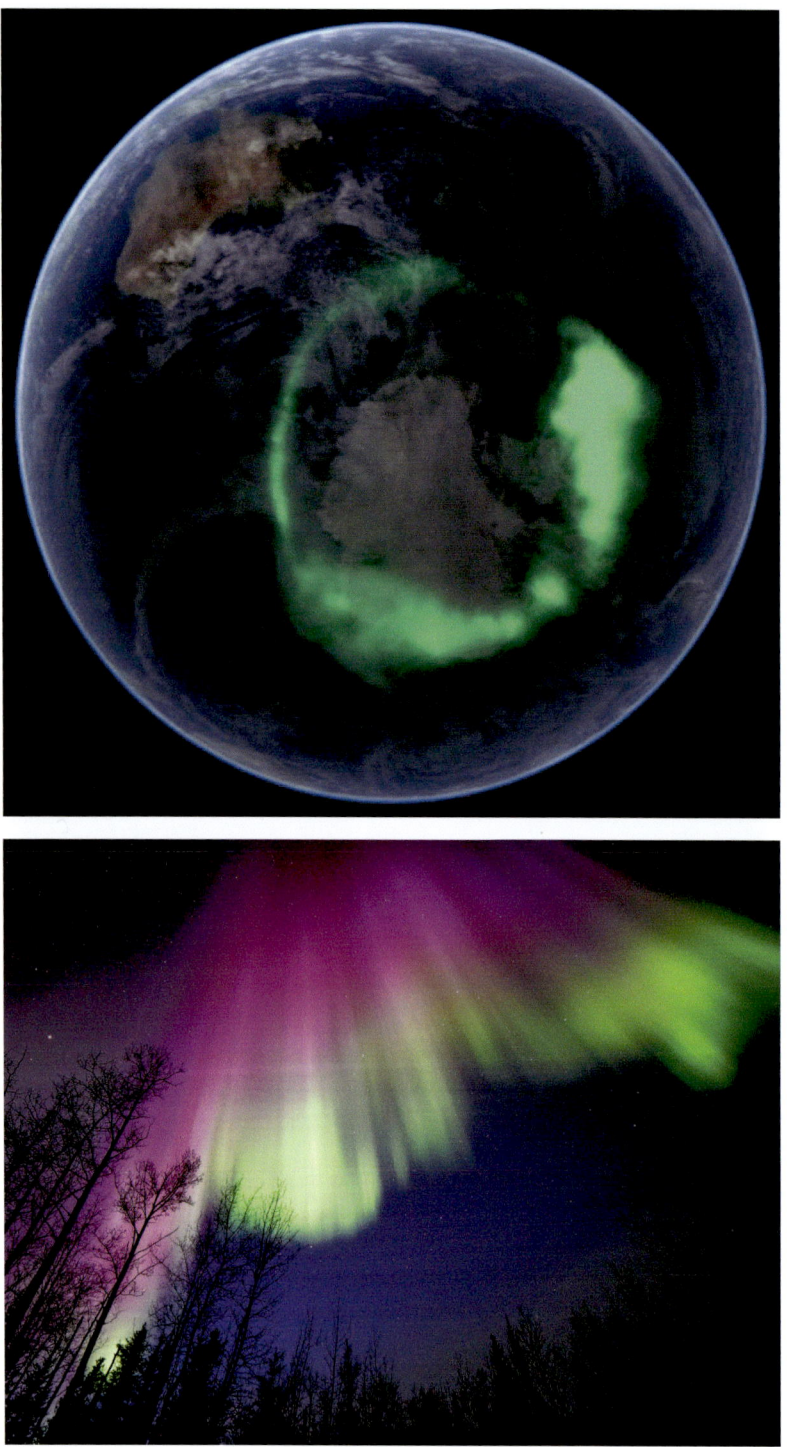

Figure 34: Aurora Borealis as seen from space (top) and from surface (bottom). Courtesy of NASA.

BOOK D FROM ΜΕΤΕΩΡΟΛΟΓΙΚΑ

Aristotle's notion on thermodynamic equilibrium

In Book D, Aristotle discusses how the two active principles (warm and cold) act and modify the two passive principles (dry and moist) and how from their interaction new substances are born and other substances decay and get destroyed. As we mentioned in the introduction, this book has hardly any relevance for meteorology. We will not discuss this book here. We will mention, however, that it appears that hidden in this discussion is the notion of (thermodynamic) equilibrium. Cold, warm, dry, and moist, are the very basic ingredients of atmospheric thermodynamics and are found behind many atmospheric processes, such as cloud and rain formation.

Concluding remarks

As we stated in the introduction the purpose of this book was not just to make the comparison between Aristotle's explanations of meteorological phenomena and today's knowledge. The real desire here was to project to the readers the brilliance of Aristotle's insights not only when he was accurately describing a phenomenon, but also when he was wrong.

Aristotle acted not only as a philosopher; he went far away from supernatural explanations and way beyond superstition. He had an unusual gift of identifying, organising, and interpreting sensory information combined with an extreme ability for deductive logic. He was also very observant and included whatever observation and information was supplied to him into his elaborate explanations. With his basis being his view of the universe, the four principles and four elements, his axioms of dry and moist exhalation, and whatever knowledge was accepted in his time, he established the standards on Meteorology that were to last for centuries and destined to shape science forever.

True enough, Aristotle came short in correctly explaining some key phenomena. For example, his dry and moist exhalation and his unawareness of atmospheric pressure, clearly prohibited him to accurately explain the genesis of winds and phenomena associated with thunderstorms, such as thunder and lightning. Also, the very important fact that Earth rotates on its axis and revolves around the Sun is absent from the Aristotelian model. We believe that if Aristotle had been able to grasp those two facts, the world would be very different today. Nevertheless, Aristotle's shortcomings teach us a lesson. Assumptions and logic may lead to explanations of phenomena, but it does not mean that the assumptions are correct. For example, big mirrors transmit images, small mirrors transmit colour, is a brilliant way to explain the rainbow, but the assumption is wrong (still, considering the general level of scientific knowledge at the time, the Aristotelian explanation of the rainbow was a great contribution to scientific knowledge). This mistake is made repeatedly even today by scientists in many important problems.

However, we have to admit that many and important explanations/descriptions of Aristotle have stood the test of time. The realisation that warm air is lighter than cold air and that warm air rises and cold air sinks, his description of convection, the fact that when the air is cooled water vapour condenses to produce clouds and rain, his paramount description of the hydrologic cycle, his insights on the formation of dew, frost, and precipitation (rain, snow, and hail), his discussion on global climate change, his arguments on the expected variability in space and in time of precipitation and droughts, the fact that most winds are north and south, the notion of rotation causing the direction of wind to be oblique, the discussion on habitable regions, and the extraordinary prediction of the existence of Antarctica, will stand the test of time forever. Aristotle moves very easily from local to global. The description of the Etesian winds and his insights on atmospheric circulation and atmospheric cells such as the Hadley cell discovered 22 centuries after him, are indeed astonishing. The water distillation, the salinity concept, the different density of different waters and its implications for a difference in buoyancy between rivers and oceans provide the profile of a great thinker and a unique observer of nature. In fact, we find him quite entertaining.

Great minds were called in ancient Greece 'Επιστήμονες', from the term 'Επιστήμη', which today means science. In its root, however, 'Επιστήμη' comes from the verb 'επίσταμαι' which translates to 'to know something very well'. Aristotle was definitely, an Επιστήμονας. He did not just know well one subject, but many.

Appendix I

Aristotle on climate change

Very few know that Aristotle acted not only as a philosopher but at the same time he went far away from supernatural explanations and beyond superstition. In fact, he was among the first scientists to try to present and hint on climate change both from a global and a regional perspective. This brief essay presents highlights of Aristotle's views on climate change.

The word 'climate' originates from the Greek verb 'kleno' (in Greek κλίνω). In the ancient thinking meant to be inclined, referring to the inclination of the sun's rays as we move from the equator to the pole, due to the spherical geometry of the Earth, which as we all know determines to a large extend climate. Aristotle considered the Earth as being a sphere at the center of the Universe and defined five climate zones based on the inclination of the sun's rays, concepts that have been put forward much earlier by Pythagoras and his student Parmenides (6th century BC). According to the knowledge of his time there were five climatic zones in a spherical world to temperate, two polar (frigid), two temperate and one torrid separating them. The torrid zone, called 'diakekavmeni zoni' was considered as being so hot that it was practically impossible to explore. The term 'tropical' came also from the Greek verb 'trepome' (in Greek τρέπομαι) which means the change in the course or the 'path' of the sun in his apparent movement from one hemisphere to the other hemisphere. The change to 'trope' (in Greek τροπή), in later centuries, gave birth to the term 'tropical' which refers to the tropical latitudes. The nomenclature of classical Greek understanding of different climates has been widely used by renaissance mappers.[1,2] The persistence in time of the Greek (Aristotle's) idea of latitudinal climatic zones provided an interesting challenge to the explorers in the renaissance who have noticed already that zonality was not the best representation of the various climates and this has been taken into account in Koeppen's classification of climates as late as in 1900.[3]

Aristotle's pupil Theophrastus has extended Aristotle's description of climatic changes on his treatise 'On winds and on weather signs' which was based on Aristotle's climatic approach and his predecessor physician Hippocrates, the first bioclimatological text ever to be written entitled: 'On Airs, Waters and Places' (5th century BC). Several Greek philosophers have quoted statements of environmental changes including climate change. In particular, Aristotle's pronouncement has a measure of relevance to the 'Mycenaean drought' (1200 BC) problem which led to an interdisciplinary controversy among some classical archaeologists and historians as well as meteorologists.[4]

With the above history being a prelude, we now move on to Aristotle's insights on climate change as described in his book Μετεωρολογικά (Meteorology) (Book A, Chapter 14):

> 'It is not always that the same regions on Earth are dry or humid, but they change according to the appearance or disappearance of rivers. This is the reason land and sea are interchangeable. Land and sea are not fixed in space, but we find sea where it used to be land and where now there is sea, it will become land again. We need to admit that these changes follow some order and periodicity. The basic principle and cause is that Earth's interior experiences a time of maturity, like the bodies of plants and animals, and a time of decay; the difference being that the changes in plants and animals do not happen in specific part of them but they necessarily mature and decay as a whole. On the contrary, on Earth the changes occur in certain places under the influence of cold and heat, which depend on the course of the Sun. For the regions becoming dryer, it is the destiny of water springs to dry out, and as such big rivers become smaller and smaller and then dry out completely. Such changes affect the sea as well. In those regions, where fueled by the full rivers, the sea flooded the land,

---

[1] Homem, L.: (Mappamonde) Lopo Homem cosmographo cauasero, Cortesao, Armando: Portugaliae Monumenta Cartographica, Lisboa, 1960. Clements Library, University of Michigan, Ann Arbor, MI., 1554
[2] Blaeu, W. J.: Nova Totius Terrarum Orbis Geographica ac Hydrographica Tabula, auct: Guiljelmo Blaeuw. Excuedebat Gulielmus Blaeuw Amsterodomi. Le Theatre du monde, Amsterdami, 1643-1646, Vol. 1, Clements Library, University of Michigan, Ann Arbor, MI., 1648
[3] Sanderson, M.: The classification of climates from Pythagoras to Koeppen. Bulletin of the American Meteorological Society, 80, 669–673, 1999.
[4] Zerefos, C.S., and Zerefos, E.: Climatic change in Mycenaean Greece: A citation to Aristotle. Arch. Met. Geoph. Biokl., Ser. B, 26, 297-303, 1978.

now it recedes leaving behind dry land. But the time will come when those places will be flooded again.'

'The physical change of Earth happens gradually and over appreciably long-time intervals compared to our length of life. As such, these phenomena pass unnoticed and whole nations are lost before they can preserve the memories of these changes from their beginning to their end. The most utter and sudden catastrophes are due to wars, pestilence, and famine. Some of these famines are of great scale, others of smaller scale. In the latter case, the disappearance of a nation may not be noticed, because some leave the country and some stay behind, until the land is unable to provide food to the few remaining inhabitants. From the time of the initial to the last departure, a long time has passed and nobody anymore remembers. The lapse of time destroys all record even for those last inhabitants that may still be alive. We also need to admit that in the same way, the date when first a nation settled in a land that was changing from wet to dry, disappears from memory. Because the change in the land occurred imperceptibly, and nobody remembers who were the first inhabitants, when they came, and what was then the state of the land.'

'This is what happened in Egypt. It is well known that this land becomes more and more dry and that the whole country is a deposit of the Nile. As the marshes gradually dried, neighboring peoples settled there but with the passing of time they forgot their origins. It is certain that all the mouths of the Nile, except that at Canopus,[5] are human made, and that in older time Egypt was called Thebes. The higher regions where inhabited earlier than the lower ones. The places closer to the silt deposit by the river, were necessarily marshy for a longer time. Subsequently, those places change and become in turn more prosperous, because as they gradually dry acquire better quality. Those places, however, where initially there was a balance between dryness and humidity, became progressively worse.'

'This also happened in the land of Argos and Mycenae in Peloponnese, Greece. At the time of the Trojan wars the land of Argos was wet and marshy and could only support a small population, whereas the land of Mycenae was thriving (which may explain its greater fame). Today the opposite is true. The land of Mycenae has dried out, but the valleys in the land of Argos are now arable. Given that the above changes can actually take place over regions of small sizes, we must admit that the same can happen over larger areas and even over whole countries.'

---

[5] The farthest to the west, it is the only natural mouth of river Nile.

'The shortsighted ones believe that this kind of events lurk in some universal change, in the sense of a coming to be of the universe as a whole. This is why they maintain that, if the volume of the sea is reduced because of dry conditions, it is because today dry conditions are occurring in more places than in the past. There is some truth to this argument, because indeed today there are more places that used to be under water. However, if they examine this issue more carefully they will observe the opposite; they will find places where the sea has invaded the land. We should not, however, think that the reason for this is a universal change. For it is absurd that for small and trifling changes on Earth, to evoke the whole universe; after all the size of Earth is negligible compared to the universe. Rather, we should attribute the reason for these changes to the fact that they occur at regular time intervals. For example, while every year there is a winter season, after some determined long time interval a great winter will come accompanied by torrential rains and flooding.[6] This great flooding does not occur always at the same place. For example, the flood in the time of Deucalion[7] affected mostly the Greek area and especially Ancient Greece (Hellas) in the vicinity of Dodoni[8] and of Achelous river,[9] which often changes its course. In this area lived the Selli and those formerly called Graeki and now Hellenes. When, therefore, heavy rains occurred, we must assume that they lasted for a long time, and whatever is happening today with the rivers (some of which flow all the time and some don't), was happening then as well. Researchers support the idea that this is because of the presence of huge underground cavities, but we attribute it to the size of the mountainous regions, to their density, and the cold that dominates them (because those regions they catch, store, and produce the larger portion of water, whereas mountainous regions of small size or porous or stony regions, see their water flowing away). We must believe, therefore, that the same is happening during great floods: In regions where water is accumulated, humidity increases thereby making them almost inexhaustible. But, as time goes on, the regions that dry out become more common, the wet regions less common, until we reach once again the beginning of the same cycle.'

'However, because some change in the universe must necessarily be happening, (without this being associated with coming to existence or with perishing; after all the universe is unchangeable), it should not be, in our opinion, that the same

---

[6] Aristotle lived in Greece where snow is hardly seen, except high in the mountains.
[7] According to the myth, Deucalion was a king in the age of copper. When Dias (Zeus) decided to eliminate him, because he was a corrupted person, Deucalion, after his father's Prometheus advise, built a boat where he loaded the necessary, and thus, he and his wife Pyrra escaped the cataclysm.
[8] In Northwestern Greece.
[9] In Western Greece.

places are always flooded or dry out. This is proven by facts. Let us consider the Egyptians again, who are considered the most ancient of people. As we mentioned previously, their whole land is obviously the making of the river Nile and this is clear to anybody who takes a look around this country. An irrefutable proof is the Red Sea. One of their kings tried to make a canal there, in order to join the Nile with the Red Sea (for it would be a great profit if all this region became navigable; it is said that Sesostris was the first of the ancient kings, who attempted it, but he found out that the Red Sea was at a level higher than the land.[10] For this reason, Sesostris first and then Darius,[11] stopped the building of the canal. They were afraid that by mixing sea and river water, the river will disappear. It would therefore appear that this part of the world was once an unbroken sea. For the same reason, in Libya, the region of Ammon[12] is lower and hollower than the land towards the sea. Clearly, silt deposits created a barrier which, as a result, formed lakes and dry land, and that with the passage of time, water in the lakes evaporated and now is gone. The same happened in lake Maeotis,[13] where deposits by rivers were so significant, that the ships now have to be smaller to sail into it. As we have argued, this lake as well, is the making of the rivers and that at some time in the future it will all dry completely. Yet another example is Bosporus,[14] where due to river deposits we always observe strong currents. These strong currents come about because every time the current from the Asiatic shore throws up a sandbank, a small lake is formed, which then dries out, then a second sandbank, and a second lake, which then dries out, and so on. After many such repetitions, is obvious that the Bosporus strait must become some kind of a river, which in the end will also dry out.'

'It is, therefore, obvious that, since time is endless and the universe is eternal, neither Tanais[15] nor the Nile have always been flowing and that the region where they flow, was some time ago dry; for their energy has limits, while time does not. This is equally true for all rivers. But if indeed, rivers form and then dry out, and if the same regions on Earth are not always covered by water, it necessarily follows that the sea is also subject to similar changes. And if the sea is receding from some

---

[10] This perception was maintained until the 19th century. During Napoleon's expedition to Egypt (1798), the level difference was estimated to tens of meters!
[11] This is Darius A who reigned over Persia and Egypt from 521 to 486 BC.
[12] Today known as Siwa oasis, 560 Km from Cairo. In Ammon, there was the famous temple of Zeus, which Alexander the Great visited.
[13] Known today as the Sea of Azov. It is bounded in the northwest by Ukraine, in the southeast by Russia. The Don River and Kuban River are the major rivers that flow into it.
[14] Bosporus is a natural strait located in northwestern Turkey. It forms part of the continental boundary between Europe and Asia. It is the world's narrowest strait and it is used for international navigation. Bosporus connects the Black Sea with the Sea of Marmara, and, via the Dardanelles to the Aegean and Mediterranean seas.
[15] Tanais was an ancient Greek city in the Don river delta.

places but it advances over other places, it is clear that over the whole area of Earth, not always the same areas are sea or land, but the picture changes with time.'

In addition to the above, in chapter 4 of Book B, Aristotle touches on the spatial and temporal variability of climate: 'Because sometimes it is the moist that dominates and sometimes it is the dry, some years are wet and some are dry and windy. It also happens, that sometimes, excessive rains or droughts to be frequent and to affect whole countries, or some regions more than others. It is also often possible that, while one region is experiencing drought conditions, regions around it get the normal amounts of rain expected in a given season. Contrariwise, it is possible that a place in the middle of a country, which normally experiences little rain or even drought, to receive large amounts of rain. The explanation for this is the following: While in most cases it is natural to observe the same phenomenon over most of the area of a country, for the reason that nearby places have a similar position with respect to the Sun (unless there is something special about them),[16] yet sometimes dry conditions will replace naturally occurring moist conditions in one area, and vice-versa, as they may move around driven by the wind.'

Clearly, Aristotle, is talking about natural climate change occurring at all space/time scales due to intrinsic variability of the system. Moreover, in statements like '*We need to admit that these changes follow some order and periodicity*' and '*On the contrary, on Earth the changes occur in certain places under the influence of cold and heat, which depend on the course of the Sun*', Aristotle, with his deductive intuition, may be making connections to what we know today astronomical forcings on climate. We note that, while the human intervention to the environment has taken place always on Earth (for example the invention of agriculture during the Neolithic era), Aristotle could not have been aware of the effect of human activity and CO_2 emissions and other human activity on the climate. Nevertheless, when it comes to *natural* climate change he was right on target. His insights on this subject, we believe, are stunning and they echo modern observations and predictions for future climate change. Changes in the frequency of many extreme weather and climate events have been observed since about the 1950s. Some of these changes have been linked to human influences, including a decrease in cold temperature extremes, an increase in warm temperature extremes, an increase in extreme high sea levels and an increase in the number of heavy precipitation events in a number of regions. As to the future, the IPCC reported estimates that the cumulative emissions of carbon dioxide shall largely determine global mean surface warming by the late 21st century and beyond.[17] Projections of greenhouse gas

---
[16] Notion of microclimate.
[17] IPCC 2014: *Climate Change 2014: Synthesis Report. Contribution of Working Groups I, II and III to the Fifth Assessment Report of the Intergovernmental Panel on Climate Change* [Core Writing Team, R.K. Pachauri and L.A. Meyer (eds.)]. IPCC, Geneva, Switzerland, 151 pp.

emissions vary over a wide range, depending on both socio-economic development and climate policy. Surface temperature is projected to rise over the 21st century under all assessed emission scenarios. It is very likely that heat waves will occur more often and last longer, and that extreme precipitation events will become more intense and frequent in many regions. The oceans will continue to warm and acidify, and global mean sea level is predicted to rise.[18] Climate change will amplify existing risks and create new risks for natural and human systems. Risks are unevenly distributed and are generally greater for disadvantaged people and communities in countries at all levels of development.

---

[18] IPCC 2012: Managing the Risks of Extreme Events and Disasters to Advance Climate Change Adaptation. A Special Report of Working Groups I and II of the Intergovernmental Panel on Climate Change [Field, C.B., V. Barros, T.F. Stocker, D. Qin, D.J. Dokken, K.L. Ebi, M.D. Mastrandrea, K.J. Mach, G.-K. Plattner, S.K. Allen, M. Tignor, and P.M. Midgley (eds.)]. Cambridge University Press, Cambridge, UK, and New York, NY, USA, 582 pp.

Appendix II

On the meaning of the word virtue and Aristotle's poem 'Ode to Virtue'

The word **virtue (in the Greek language αρετή)** presents a concept praised by Hellenism from Homer to Aristotle who composed the best hymn to virtue cf. in Croiset [*History of Ancient Greek Literature*, Papyrus Publishing, Athens, 1938]) of which only the beginning has survived to us. Nor is anyone able to understand the culture of Ancient Greece without the full comprehension of this word, to the definition of which Plato dedicates an entire dialogue, "Protagoras", while he constantly returns to this word in almost all his other dialogues. And the virtues which are related to religion, the republic, philosophy, mythology, the Eleusinian Mysteries, are the following seven: the natural virtues; the moral; the political; the purgatory; the theoretical; the exemplary; and lastly, the virtues that are superior to all others, the priestly or theurgical or royal virtues, which, along with the three previous ones, constitute the so-called divine virtues. Plato was the first to define both the essence of virtue and its four components (wisdom, courage, justice, prudence); Aristotle ensued, with his "Nicomachean Ethics", the most mature of his works, being one of the most magnificent monuments ever created by the human mind as it regards to virtue. Porphyry has also dealt with virtue scientifically and most seriously, following the ancient Greek tradition, mainly in his treatise titled «Αφορμαί εις τα νοητά» (*Sententiae Ad Intelligibilia Ducentes*). Last but not least, the one who has most analytically saved what is known about the 7 aforementioned virtues, is Olympiodorus, so little known, yet whose works contain an invaluable treasure of knowledge and revelation, more specifically in his commentary on Plato's "Phaedo", while he certainly is aware of all the Neoplatonic philosophers that preceded him, and especially Proclus. Moreover, these 7 virtues, as we have already mentioned, constitute the gold key to understanding the Greek religion, mythology, philosophy, Eleusinian mysteries and science. Based on the Greek edition of "Πάπυρος" Publishing Company, the fragment of Aristotle ode or hymn to virtue was published in Greek (ancient and modern Greek by Andreas I. Pournaras, "Πάπυρος" Publ. Co., 1938). These texts (in Greek) read as follows:

Original Ode to Virtue by Aristotle in Greek (only the beginning has survived)

Ἀρετά, πολύμοχθε γένει βροτείῳ,
θήραμα κάλλιστον βίῳ,

σᾶς πέρι, παρθένε, μορφᾶς

καὶ θανεῖν ζηλωτὸς ἐν Ἑλλάδι πότμος

καὶ πόνους τλῆναι μαλεροὺς ἀκάμαντας·
τοῖον ἐπὶ φρένα βάλλεις

καρπὸν, ἰσαθάνατον χρυσοῦ τε κρεῖσσω
καὶ γονέων μαλακαυγήτοιό θ' ὕπνου.

σεῦ δ᾽ ἔνεχ᾽ οὐκ Διὸς Ἡρακλέης Λήδας τε Κοῦροι

πόλλ᾽ ἀνέτλασαν ἔργοις
σὰν ἀγρεύοντες δύναμιν.
Σοῖς δὲ πόθοις Ἀχιλεὺς Αἴας τ᾽ Ἀΐδαο δόμους ἦλθον·

σᾶς δ᾽ ἕνεκεν φιλίου μορφᾶς Ἀταρνέος

ἔντροφος ἀελίου χήρωσεν αὐγάς.

τοιγὰρ ἀοίδιμος ἔργοις,

ἀθάνατόν τε μιν αὐξήσουσι Μοῦσαι,

Μνημοσύνας θύγατρες, Διὸς ξενίου σέβας ἀσκουσαι

φιλίας τε γέρας βεβαίου.

Translation in modern Greek by Andreas I. Pournaras
(Πάπυρος, Athens, 1938)

Ω Αρετή, αντικείμενον των μόχθων του γένους των θνητών, ωραιότατον θήραμα του ανθρωπίνου βίου, προς χάριν σου, ω παρθένε, θεωρείται επίφθονος τύχη και αυτός ο θάνατος εις τους Έλληνας και αυτοί οι σκληροί άνευ αναπαύσεως κόποι, τόσην αγάπην προς σε εμβάλλεις εις τας καρδίας αθάνατον και ανωτέραν από τον χρυσόν και από τους γονείς και από τον καταπραυντικόν ύπνον. Δία σε, ο υιός του Διός Ηρακλής και τα δίδυμα της Λήδας υπέφεραν μυρία κακά επιδιώκοντες την δύναμίν σου. Δι' αγάπην προς σε, ο Αχιλλεύς και ο Αίας κατήλθον εις τον Άδην. Διά την προσφιλή σου μορφήν, το θρέμμα των Αταρνών (ο Ερμείας) εγκατέλειψε το φως του ηλίου. Δι'αυτό αι πράξεις είναι άξιαι ασμάτων και θα του δώσουν την αθανασίαν αι Μούσαι, αι θυγατέρες της Μνημοσύνης, χάρις εις τας οποίας θα διαλάμψουν προς τιμήν του η λαμπρότης του ξενίου Δία και τα δώρα της πιστής φιλίας.

Translation in English by C.D. Yonge

O Virtue, won by earnest strife,
And holding out the noblest prize
That ever gilded earthly life,
Or drew it on to seek the skies;
For thee what son of Greece would not
Deem it an enviable lot,
To live the life, to die the death
That fears no weary hour, shrinks from no fiery breath?

Such fruit hast thou of heavenly bloom,
A lure more rich than golden heap,
More tempting than the joys of home,
More bland than spell of soft-eyed sleep.
For thee Alcides, son of Jove,
And the twin boys of Leda strove,
With patient toil and sinewy might,
Thy glorious prize to grasp, to reach thy lofty height.

Achilles, Ajax, for thy love
Descended to the realms of night;
Atarneus' King thy vision drove,
To quit for aye the glad sun-light,
Therefore, to memory's daughters dear,
His deathless name, his pure career,
Live shrined in song, and link'd with awe,
The awe of Xenian Jove, and faithful friendship's law.

Index

aether 10-11, 13
Anaxagoras 13, 39, 54, 75
Anaximander 9
antiperistasis 55
Archimedes 63
Aristarchus of Samos 18
Aristotle on
    Antarctica 74
    Aurora Borealis 100
    Buoyancy 63
    climate change 56-60, 111-117
    clouds 13-16, 36
    dew and frost formation 37-38
    direction of winds 65-67
    etesian winds 69
    hail formation 38-39
    halo formation 85, 89-90
    inhabitable regions 70, 72-73
    rain formation 16-17, 38, 64
    rainbow formation 85-86, 92-93
    secondary rainbow 97-98
    snow formation 38
    spatial and temporal variability of rain and droughts 66
    sun dogs and light pillars 99
    thunder and lightning 74-75
        types of precipitation 38
        water distillation and salinity 63
    winds 63-64, 69
Aristotle's
    conjecture of Antarctica 74
    cyclic motion of the Sun 16
    four elements 11-12
    four principles 11-12
    hydrological cycle 16, 36
    law of reflection 96-97
    model of the universe 9-10, 15
    notion of convection 16
    notion of exhalation 14-15
    notion of microclimate 66, 116
    theory of color 86-89
axioms 64
Copernicus 19
Empedocles 11, 75
epicycles 18
exhalation 14-15, 64-68, 74-75
Galileo 19
heliocentric model 18
Herodotus 16
Hippocrates 63
Oceanus 16, 36-37
Plato 9, 12
Ptolemy 18
adiabatic cooling 31

angular momentum 79
antisolar point 95
atmospheric pressure 21
Aurora Borealis 99-101
Avogadro 24
capacity (for water vapor) 41
cold front 66
composition of the atmosphere 20-21
conduction 26
convection 26
Coriolis force 67
cumulonimbus cloud 75
equilibrium vapor pressure 43-44
etesian winds 69-71
formation of halo 90-91
formation of hail 50-51
formation of light pillars 98-99
formation of rainbow 93-95
formation of secondary rainbow 96
formation of sundogs 98-99
formation of thunder and lightning 75-79
formation of thunderstorm 75
formation of tornadoes 79-80
Hadley cell circulation 29
homosphere 20
heterosphere 20
humidity 41
hydrostatic balance 21
ice crystals 47-49
ideal gas law 24
mesocyclone 79
mole (definition) 25
latent heats 24
low-pressure systems 65-66
mesosphere 20, 33
Milenkovitch 60
Molecular weight of dry air 26
Molecular weight of moist air 26
Mpemba effect 55
parcel of air 30
pressure gradient force 21
rain formation
        warm cloud 45-46
        cold cloud (ice crystal process) 47-49
reflection of light 91, 93-95
refraction of light 93-95
relative humidity 41
snow formation 49
stratosphere 20, 33
thermosphere 20, 33
troposphere 20, 33
vertical structure of the atmosphere 33
virtue (ode) i, 5, 117–119
warm front 66
wind 21